IRRIGATION IN SOUTHERN PERU: THE CHILI BASIN

by
Roland C. Holmes
University of Southern Florida

THE UNIVERSITY OF CHICAGO
DEPARTMENT OF GEOGRAPHY
RESEARCH PAPER NO. 212

1986

Published 1986 by The Department of Geography
The University of Chicago, Chicago, Illinois

Library of Congress Cataloging-in-Publication Data

Holmes, Roland C. (Roland Clifford), 1929-
Irrigation in southern Peru.

(Research paper / University of Chicago, Dept. of Geography ; no. 212)
Bibliography: p. 193
1. Irrigation farming--Peru--Chili River Watershed. 2. Irrigation--Peru--Chili River Watershed. 3. Agricultural geography--Peru--Chili River Watershed. I. Title. II. Series: Research paper (University of Chicago. Dept. of Geography) ; no. 212.
H31.C514 no. 212 910 s 86-6931
[S616.P4] [338.1'62]
ISBN 0-89065-117-5

Research Papers are available from:
The University of Chicago
The Department of Geography
5828 S. University Avenue
Chicago, Illinois 60637-1583
Price: $10.00; $7.50 series subscriptions

CONTENTS

LIST OF TABLES

LIST OF ILLUSTRATIONS

ACKNOWLEDGMENTS

Many people in Peru and the United States have aided the author in bringing this study to completion. Much credit goes to the residents of the Department of Arequipa including farmers and governmental officials. Particular thanks goes to Professor Hermán Ugarte Chamorro of the Universidad Nacional de San Agustín for his invaluable help in many ways.

The author is indebted to the Fulbright Commission for the award of a scholarship to study in Peru. This study owes a great deal to Gilbert White. The author expresses his gratitude to him for his patient guidance of this study. Particular thanks go to Professor Norton Ginsburg for his review of a draft of the manuscript. The maps and diagrams are the work of Joanna Martin and Kenneth Kuhar.

CHAPTER I

MANAGING IRRIGATION WATER

The future expansion of irrigated agriculture in Peru raises issues of public policy whose resolution will depend in part on thoughtful examination of conditions affecting the human use of water.[1] A productive irrigation scheme reflects the effective management of water, the single most important natural resource in irrigation agriculture.[2] Two critical issues that have not received adequate attention in the development of irrigation in Peru are why farmers manage water as they do at the farm level in the face of technical knowledge suggesting greater productivity by alternative methods, and how improvement in the use of this resource may be promoted in new irrigation projects.[3]

The purpose of this study is to examine these two critical issues by looking at how both farmers and "technico" experts perceive and manage irrigation water in an area representative of some Peruvian coastal drainage basins. In assessing this situation, the specific aim is to inquire into the following questions:

1. How do irrigators in the study area differ from each other in their management of water?

---

[1] For an elaborate review of potential irrigation projects in Peru, see República del Perú, Ministerio de Agricultura, Oficina Sectorial del Planificación Agraria, Inventario de Proyectos para Incrementar en un Millión de Hectares; El Area Cultivada con Programa Preliminar de Ejecución y Estimación de Inversiones (Lima: Editorial Servicio de Copias "Fast," 1965).

[2] In a similar vein "Geertz argues that it is the supply and control of water rather than the type of soil, which is the key variable in wet-rice cultivation . . . ." Rhoda Helperin, "Conclusion: A Substantive Approach to Peasant Livelihood," in Peasant Livelihood, Studies in Economic Anthropology and Cultural Ecology, ed. Rhoda Halperin and James Dow (New York: St. Martin's Press, 1977, p. 273.

[3] A recent comprehensive review of problems and issues confronting irrigated agriculture in the developing areas of the world is E. Barton Worthington, ed., Arid Land Irrigation in Developing Countries: Environmental Problems and Effects (New York: Pergamon Press, 1977). See especially section I, "The Main Effects and Problems of Irrigation," pp. 3-72. For a comprehensive review of the characteristics, problems, and potential for development of the arid lands of South America in general and of the Peruvian desert in particular, see: Arturo Cornejo T. , "Resources of Arid South America," in Arid Lands in

2. How do they depart from what the "technicos" regard as optimal?
3. How do we account for these two sets of differences?

The Significance of the Study

The use of irrigation water is at the heart of Peruvian agriculture, and agriculture is vital to the nation's economy. Not only is over half the population dependent upon agriculture, but it is also one of the country's principal sources of foreign exchange.[1] The entire cultivated area along the coast, to which irrigation is essential, represents 25 percent of the total cultivable land and 60 percent of the value of the total agricultural production.[2]

Improvement in the use of the present supply of irrigation water is most important for Peru in two ways. First, the amount of water saved by conservation of existing water supplies would permit the irrigation of an additional area equivalent to that presently irrigated.[3] Thus, new irrigated land would be available without the need to construct additional costly dams and reservoir sites.

A second benefit Peru may derive from improving the use of the present supply of irrigation water can have immediate economic results. Coutu and King,[4] in suggesting the need for increasing agricultural production within a short time, believe that improvement in the use of existing cultivated land, including water management, is an important means to this goal.

Transition, ed. Harold E. Dregne, Publication no. 90 (Washington, D.C.: American Association for the Advancement of Science, 1970), pp. 345-380.

[1] United Nations, Food and Agriculture Organization and the International Bank for Reconstruction and Development, The Agricultural Development of Peru, Part I: General Report (Washington, D.C.: IBRD, 1959), p. 1; U.S., Department of Commerce, Overseas Business Reports, Basic Data on the Economy of Peru, prepared by B.B. Sever (Washington, D.C.: Government Printing Office, December 1969), pp. 5 and 8.

[2] Comité Peruana de Zonas Aridas, Informe Nacional sobre las Zonas Aridas, Para la Conferencia Latinoamericana sobre el Estudio de las Regiones Aridas en Buenos Aires (Lima: Imprenta Equiduplic, 1963), p. 103; Arthur J. Coutu and Richard A. King, The Agricultural Development of Peru (New York: Praeger Publishers, 1969), p. 31.

[3] United Nations, Food and Agriculture Organization and the International Bank for Reconstruction and Development, The Agricultural Development of Peru, Part II: Detailed Report, Water Utilization, p. 135. For a general discussion on the importance of improving existing irrigation projects as compared with the establishment of new projects, see V.A. Kovda, ed., Irrigation, Drainage, and Salinity: An International Source Book (FAO UNESCO) (London: Hutchinson and Company, 1973), pp. 482-483.

[4] Coutu, Agricultural Development of Peru, pp. 120, 123, and 135.

From the economic expert's point of view, present techniques related to farm management of irrigation water have not reached optimal levels as defined by the "technicos" trained in engineering.[1] Unfortunately, achieving improved on-farm water management appears to be one of the more difficult of the development problems in irrigated agriculture. In attempting improvements, an ever-popular solution is the introduction of high-level irrigation technology.

In the past, there has been some effort in Peru to better the physical efficient use of water,[2] but with little success. Different approaches were taken toward irrigation by farmers and government officials. Having taken a normative approach to the planning of irrigation schemes, the government has not utilized as much information as it could have about the success or failure of past projects. The authorities seem largely to have ignored the reasons that farmers follow prevailing practices.

Geographers, students of the transfer of water resource knowledge, and those concerned specifically with the development of irrigated agriculture, may find this study significant because the situation in the area clearly shows the disjointed progression from old to new irrigation schemes, with little evidence of change in water use practices, particularly at the farm level. This study may be noteworthy also because it depicts the nature of daily water management in irrigated agriculture in existing projects.

Interviews with some of the agricultural leaders in Peru give one the impression that, although a number of them note the importance of the management of irrigation water at the farm level, few, if any, are willing to give it top priority in agricultural planning. In fact, in developing new irrigation schemes, the design engineer and the managing directors of the colonizaiton process are rarely concerned about learning how farmers in nearby irrigated areas manage their water and so are not fully aware of the milieu in which they as technicians are working.[3]

---

[1]Comité Peruana de las Zonas Aridas, Informe Nacional sobre las Zonas Aridas, p. 51; Inter-American Committee for Agricultural Development, Inventory of Information Basic to the Planning of Agricultural Development in Latin America: Peru (Washington, D.C.: Pan American Union, 1964), p. 51.

[2]Early efforts toward improving the physical efficient use of water in Peru were advocated by C.A. Lizarrage in "El Estudio del Riego," Aguas e Irrigación 1 (June 1932): 41-47.

[3]An example of this point of view is reflected in the initial planning by an international development agency during the decade of the 1960s in one of the north-central coastal valleys of Peru. One of this agency's intentions was to introduce sprinkler irrigation to the farmers presently using gravity methods without learning why one farmer of the valley had already abandoned a preexisting sprinkler system.

Of necessity, the irrigation schemes must be implemented for economic reasons in the shortest time possible; yet in this process some phases of colonization, such as on-farm water management, are being given short shrift, which results in the perpetuation of traditional practices. This, in turn, leads to an undue strain on water resources. Review and assessment of the use of irrigation water in the Arequipa area (Chili Basin) within the Quilca River Basin of Southern Peru will help to clarify these issues (figure 1).

Related Literature

In the social sciences there has been ample work done towards improving water resources management policy.[1] The major contributions have come from economics, with an emphasis on decision rules. An example is the digital computer model of irrigation systems developed by Anderson and Maass.[2] This model indicated the effect of water supply and operating rules on production and income on irrigated farms. It makes the assumption that irrigators have already made proper adjustments to the engineering features of the system. It is precisely this assumption that characterizes many of the normative-type studies made by both social scientists and engineers in grappling with problems in irrigated agriculture.

Geographers in a somewhat more inclusive manner,[3] within the man-land tradition of their discipline, have pushed for the consideration of a wider range of alternatives, taking into consideration the perception of resource managers in gauging how they arrive at a final choice. Much of the recent work done by geographers along these lines is in the study of natural hazards which focuses on man's perception of his environment (hazard), what alternatives are open to him, and how he goes about making a final choice in adjusting to this special environment.[4]

---

[1] L. Douglas James, ed., Man and Water (Lexington, Kentucky: The University Press of Kentucky, 1974).

[2] Raymond L. Anderson and Arthur Maass, A Simulation of Irrigation Systems U.S. Department of Agriculture Economic Research Service, Technical Bulletin no. 1431 (Washington, D.C.: Government Printing Office, 1974).

[3] As some would say, "The modern geographer is a mysterious person to one with little formal training in that subject. Geographers seem to be economists, sociologists, or anthropologists wearing false moustaches." James, Man and Water, p. 224.

[4] Ian Burton, Robert Kates, and Gilbert F. White, The Human Ecology of Extreme Geophysical Events (Toronto: University of Toronto, Department of Geography, Working Paper no. 1, 1968).

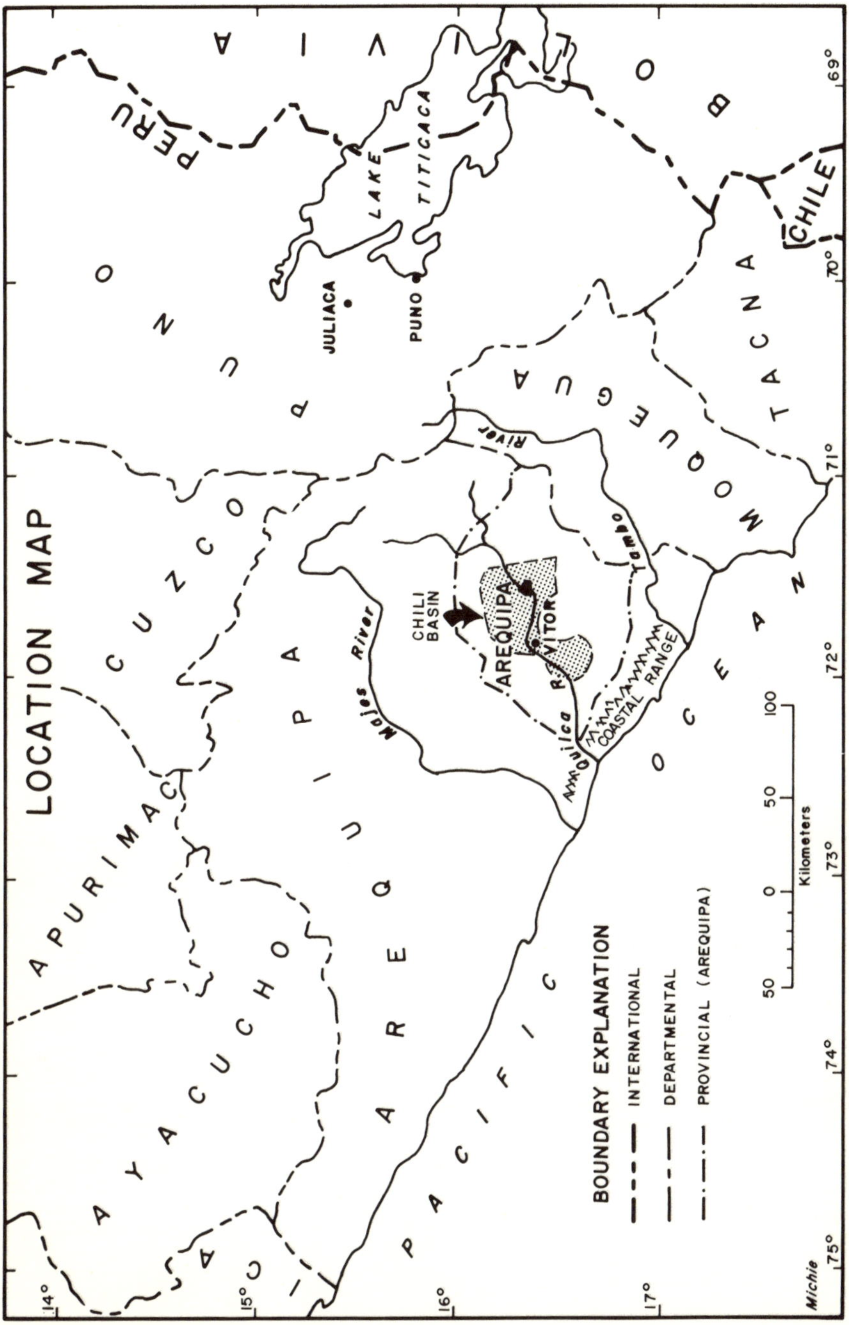

Fig. I - General Location, Chili Basin.

Also within the man-land tradition of geography, a number of elaborate studies of irrigated agriculture have been conducted. During modern times, the French geographer Brunhes examined the similarities and differences that exist between various irrigated areas of North Africa and Spain.[1] Among his most important conclusions is the prominence he attributes to the "psychological effect as the middle term between facts of the physical order and economic facts."[2] One way that he shows this is by community regulation being stringently enforced in seeking a normal and peaceful situation to a water supply of great variability. This is not determined directly by natural conditions, but is the result of a state of mind reflecting these conditions.

Another more recent study of an irrigated area is Hudon's assessment of factors that influence irrigation water use in the Utah Valley.[3] His conclusions and recommendations are evaluated as they relate to those factors which block adoption of efficient water management practices. Two specific factors are the lack of knowledge and interest on the part of the farmer and the type of institutional controls in existence. The present study amplifies Hudson's approach by examining irrigation water management in different environments as well as raising the whole spectrum of the experts' goals as compared to the farmers' in bringing about an optimal level in water management in a developing area of the world.

Hawley, in turn, in studying the use of ground water in the South Plains of Texas, addressed the question, "Are irrigators wilfully wasteful and extravagant in using ground?"[4] Some of the hypotheses he tested in analyzing the decision-making process in irrigation and which the present study assesses with modifications are:

1. Cotton acreage is irrigated when water is needed, and need is determined by local field conditions
2. Irrigation of cotton acreage is unrelated to highly efficient water distribution techniques or water conservation measures
3. The land tenure status of the farm operator determines whether or not the cotton acreage will be irrigated and the particular water use practices applied.

---

[1] Jean Brunhes, L'Irrigation dans la Péninsule Iberique et dans l'Afrique du Nord (Paris, 1902).

[2] Jean Brunhes, Human Geography (Chicago: Rand McNally, 1920), p. 536.

[3] James Hudson, Irrigation Water Use in the Utah Valley, Utah, Research Paper no. 79 (Chicago: University of Chicago, Department of Geography, 1962).

[4] Arthur J. Hawley, "The Decision to Irrigate: Irrigation Practices and the Water Supply in the South Plains of Texas" (unpublished Ph.D. dissertation, Department of Geography, University of Michigan, 1968).

An important topic related to the interactions between man and his physical world, which has been developed by geographers and others in recent years, is an understanding of the behavior of professional environmental decision-makers.[1] The relevance of this topic to the present study can be judged by indicating the conclusions of Sewell's study, as seen by Sims and Baumann.

> What Sewell superbly demonstrated in this study is... (that) when society turns to an expert for advice, it always gets more than it bargains for. In addition to his professional expertise, it gets his professional biases; along with his skills come his prejudices, of which he, himself, is, by the way, quite unaware ... Sewell shows that membership in ... (particular) professions means that ... only certain solutions are considered and proposed, and that only certain attitudes are taken toward the government and toward the public. In other words, the professionalization processes that have produced the expert's skills have, at the same time, limited his vision.[2]

There have been a number of studies to determine the successes and failures of irrigation schemes as instruments of government policy.[3] Farmer, in analyzing government policy in the development of irrigation projects in the Dry Zone of Ceylon (Sri Lanka), noted that failure in persuading farmers to adopt new farming practices including manuring, transplanting of young rice plants, weeding, and the use of the Burmese harrow.[4] Two conclusions drawn from his review of these technical problems are: (1) the colonist takes most kindly to improvements which can be most readily grafted on to his traditional methods, e.g., pure-line seed and the light iron plough, and (2) the colonist with a relatively large farm for paddy, together

---

[1]Kenneth H. Craik, "The Environmental Dispositions of Environmental Decision-Makers," The Annals of the American Academy of Political and Social Science 389 (May 1970): 87-94; W.R. Derrick Sewell, "Environmental Perceptions and Attitudes of Engineers and Public Health Officials," Environment and Behavior 3 (March 1971): 23-59.

[2]John H. Sims and Duane D. Baumann, eds., Human Behavior and the Environment (Chicago: Maaroufa Press, 1974), p. 180.

[3]B.H. Farmer, Pioneer Peasant Colonization in Ceylon (London: Oxford University Press, 1957); Arthur Gaitskell, Gezira: A Story of Development in the Sudan (London: Faber and Faber, 1959); Wolf Roder, The Sabi Valley Irrigation Projects, Research Paper no. 99 (Chicago: University of Chicago, Department of Geography, 1965); John C. deWilde, Experiences with Agricultural Development in Tropical Africa (Baltimore: The Johns Hopkins University Press, 1967); Robert Chambers, Settlement Schemes in Tropical Africa (New York: Praeger Publishers, 1969); Arthur Maass and Raymond L. Anderson, . . . and the Desert Shall Rejoice: Conflict, Growth, and Justice in Arid Environments (Cambridge, Massachusetts: The MIT Press, 1978).

[4]Farmer, Pioneer Peasant Colonization in Ceylon.

with a lack of labor at critical times, finds himself confronted with a condition that makes for extensive rather than more intensive farming methods. As a result, many of these methods have been given up by new settlers who in turn have reverted to their traditional ways.

Farmer surmised that "the caution with which the colonist, . . . views these things has often in it a perfectly rational objection to methods which he knows to be of doubtful utility. . . ."[1] He also stated that "resistance to technical change has a rational basis, as when, for example, the proposed new technique is untried, or even unusable."[2] It was pointed out that what may happen is that the colonist may merely go through the motions of conforming to the technical change.

In Farmer's study two other factors that help to explain the resistance to technical and economic change are social attributes and attitudes toward the government. Among the several attributes of Sinhalese society that help to produce resistance to change are:

1. "The superior attitude adopted towards the colonist by Colonization Officers and others whose job it is to teach new methods."[3]
2. "The colonist's attitude to wealth. If . . . [the colonists] can reach standards of consumption which they have in mind without the additional care and effort which are required by technical change, then they continue to use the old methods."[4] In addition, capitalist economic motivation is unknown.
3. "Work . . . [is] seen only as one of a number of necessary activities, a mere means of acquiring 'concrete needs determined by custom' [; therefore] there is no overriding pride in workmanship for its own sake."[5]

The colonist tends to look on the government as the giver of all things, with no effort on his part being necessary. A possible cause advanced for this attitude relates to the hierarchical organization of society and autocratic rule, which required little initiative and no participation in central government from the peasant.

Farmer believed that to be effective, the means that are used (such as agricultural education) to meet the problems of technological change should be based on indigenous cultural motives.

Whereas the project that Farmer investigated was unsuccessful in implementing government policy, the Gezira Project as related by Gaitskell[6] reflects a positive outcome. Gaitskell believed that, to be a success, an irrigation project must entail democratic participation by the people in the whole process. Even though there is

---

[1]Ibid., p. 354. [2]Ibid., p. 284. [3]Ibid., p. 285. [4]Ibid., p. 286.

[5]Ibid., p. 287. [6] Gaitskell, <u>Gezira</u>.

difficulty avoiding an autocratic attitude, the administration must change from its position of "paternal authority" to the humbler role of adviser.

Additional conditions responsible for the successful outcome of this project are:

1. The site of the project as related to the availability of water, soils, and dam site
2. The climate, which permitted a new cash crop, long-staple cotton
3. The pilot schemes, which resulted in suitable selection of crops, workable arrangement with the people, education of field staff, and better human relations between farmers and officials
4. The type of imposed cooperation, which eliminated the weakness of a farmer having to strive alone by prescribed farm procedures designed not to suppress but to support the individual
5. The method of giving help, which resulted in readily available loans for each specific task. These loans were tailored to meet immediate needs, so that a farmer's horizon of effort would not be extended far in time or scope, allowing him to adapt to a new routine
6. The nature of the field staff, which gave immediate access of staff to farmers, encouraged farmers against inertia, and assured the continuity of staff in the same area.

An additional observation made by Gaitskell, which has relevance for the present study, is the introduction of a new irrigation technique through gradual procedures. In order to avoid night irrigation, which would give farmers not yet familiar with irrigated agriculture some difficulty, the staff avoided this practice by storing night water in canals. Then, after the farmers had become familiar with this new way of farming (irrigation), night irrigation was introduced.

Anthropologists, in studying how man uses his water resources, do not suffer from the naive oversimplification that society can have a commonality of goals.[1] To them there is no universal water management policy. Applied anthropology has analyzed mistakes made by development officers in developing irrigation projects. One of the reasons given for failures relates to the assumption that experts know what the people want. This assumption ignores the views of local people who possess intimate knowledge of the present environment and resource management techniques.[2] In reviewing the

---

[1] James, Man and Water, p. 202.

[2] Ibid., p. 63. An example of engineers isolating themselves from the needs of the ultimate user, the farmer, occurred in the Indian subcontinent in the 1950s, where:

> The measurements and calculations of the engineers--in terms of stress, of needed materials, of on-site management, of maximum and minimum power potential--are all precise, exact and excellent. But, thereafter, too

conventional paradigm of development practitioners Ingersoll observed that:

> From the late 1940's until the early 1970's, the dominant effort of development agencies was to improve the economic growth of poor countries, achieved through transfer of technology and creating of infrastructure, and measured in G.N.P. . . . This massive, complex material task was basically one of action, in the form of projects. . . . The institutional values and career rewards that emerged from this tradition of development assistance . . . supported meeting bureaucratic goals within bureaucratic deadlines. Local "customs" or "superstitions" might occasionally interfere with the process, but the paradigm had no place for real consideration of social systems undergoing basic change. Instead, the emphasis was, and is, on precise measurement of quantitative variables as indicators of material progress.[1]

Palmer in discussing resettlment schemes stated that:

> . . . most of the studies by anthropologists concern the role of traditions which the settlers bring to the resettlement scheme. The anthropologists typically describe, or intend to describe, the way in which the traditional cultures of the settlers either help them to adjust or create points of disagreement with the scheme administrators. These points of disagreement are regarded as problems either for the administrator or for the settler, depending upon the role and sympathies of the anthropologists.[2]

Anthropologists in trying to introduce "social aspects" into development project operations:

> are usually called in at a late stage in project formulation--usually a stage known as appraisal, when the format and goals of the project have reached an advanced stage of

---

much has tended to be guesswork . . . Agriculturalists were not drawn into the planning of the big dams. The installations would be completed and main channels built by the well-trained and confident corps of hydraulic engineers before local ditches had been put in, farm-to-market roads constructed, storage built, soils tested, or crop patterns and varieties discussed. Water would be standing by while peasants continued dry farming. (Barbara Ward and René Dubos, Only One Earth: The Care and Maintenance of a Small Planet [New York: W.W. Norton and Company, 1972], p. 164.)

In a discussion of an adaptive approach to environmental impact assessment and management, it was noted that ". . . even the most well-meaning and self critical experts tend to be bound to their own specialties and techniques." (C.S. Holling, ed., Adaptive Environmental Assessment and Management [Chicester, England: John Wiley and Sons, 1978], p. 119.)

[1]Jasper Ingersoll, "Anthropologists and the Agency for Internation Development (A.I.D.): An Old Hate Relationship and a New Love Affair," Anthropological Quarterly 50 (October 1977): 200. (Emphasis added.)

[2]Gary Palmer, "The Ecology of Resettlement Schemes," Human Organization 33 (Fall 1974): 242.

> preparation and are crystallized. At this stage it requires even greater skill to redirect both project goals and format.[1]

While it is reare for social scientists, especially anthropologists, to participate in development projects and have a positive effect on the outcome, this is exactly what happened in the Vicos Project.[2] A group of scientists headed by Holmberg, an anthropologist, "undertook a systematic program of research and development in order to determine how an Indian population would respond to a concerted effort to introduce it to a more modern way of life."[3]

Through the efforts of Holmberg and others, this group of Indians, formerly serfs on the Vicos Hacienda in the Callejón de Huaylas of North Central Peru, in time became owners of this property. The value of this project for development officers lies in the strategy followed in acquiring and using knowledge. Holmberg's success stemmed from his ability to transfer his patronal decision-making power to Indian leaders. The beginning of this process was in the successful increase in potato production, a resultant of the practical transfer of new knowledge to adults, who made decisions about their own farming practices.

Hypotheses

Several hypotheses were formulated to guide this study in assessing the use made of irrigation water. The main hypothesis is that the management of irrigation water is in accordance with the traditions of the resource manager, be he farmer or expert. Several auxiliary hypotheses are:

1. Use of irrigation water is clearly affected by irrigation methods employed
2. Crops are irrigated when water is needed, and need is determined by local field conditions
3. The land tenure status of the farm operator does not normally determine the particular water use practices applied
4. The goals of optimal water management by irrigation experts are not realistic.

Framework

To provide the framework of the present study, a

---

[1] Raymond Noronha, "The Anthropologist as Practioner: The Field of Development AID," Anthropological Quarterly 50 (October 1977): 212.

[2] Henry F. Dobyns, Paul L. Doughty, and Harold D. Lasswell, eds., Peasants, Power, and Applied Social Change: Vicos as a Model (Beverly Hills: Sage Publications, 1971).

[3] Richard N. Adams et al., Social Change in Latin America Today: Its Implications for United State Policy (New York: Vintage Books, 1960), p. 79.

decision-making model developed by Kolars is used.[1] On the basis of a distinction between two conditions under which decision-making occurs, "conceptualized systems of farming can be made as points of evaluation and decision-making joined by chains of committed action based on crop characteristics."[2]

One set of inquiries by the resource manager relates to decision-making under conditions of commitment whereas the other set relates to those decisions under conditions of evaluation. Kolars gives examples of each set. "Is the soil in my small peanut field dry enough to irrigate; and after I irrigate how soon should I hoe?" represent the sequential and committed set; while, "Is the climate here good enough for sugar beets; is there sufficient water for irrigation; is the refinery close enough; will the price be right?"[3] pertain to the second set of conditions of evaluation that cen be considered simultaneously and in varying orders of priority.

Making this distinction between decision-making under different conditions allows a more meaningful evaluation. The example Kolars gives is "a community harassed by harvest chores will be less receptive to external attempts to bring about systematic change than if it were approached at some more opportune time."[4]

The framework will be used in assessing particularly that part of the system of farming that relates more directly to the use of irrigation water. In the evaluation, special attention will be given to such variables as differences in landscape and in managerial perception of environment and opportunities. As Kolars points out, this model will make it "easier to discover key points within the total system when changes introduced from outside may have the best chance of acceptance and success."[5]

Methodology

Source of Data

The inception of this study originated with a field study undertaken in the early 1960s in Colorado, which compared irrigation water management near Greeley, Alamosa, and Durango on the Southern Ute Indian Reservation. Using this experience as a springboard, an appraisal of the methods of land and water management in the Arequipa

---

[1] John F. Kolars, "Decision and Commitment in Turkish Agriculture," Review of the Geographical Institute, University of Istanbul, International Edition, Separate, no. 11 (1965-1968).

[2] Ibid., p. 39. [3] Ibid. [4] Ibid. [5] Ibid.

area of southern Peru was undertaken. This area was selected because it contained irrigated agriculture in remarkably varying environments and included pre-Columbian as well as modern irrigation schemes.

The study is based upon examination of irrigation projects located in different physical environments that represent a spread in age and governmental participation. The older project, which reflects the least on-going government participation, was selected from the immediate vicinity of the city of Arequipa (Campiña of Arequipa). The more recent government-sponsored schemes selected are located on the Pampa of La Joya. Here two projects are analyzed: the original La Joya Project (La Joya Antigua) established in the early 1940s and its extension (San Isidro), first colonized by farmers in 1967.

There were two investigations. The first developed a set of data which compares farm practices on several farms, the second traced the intensive effort made by the technical-scientific community to induce farmers to change their water management practices in the colonization of the new irrigation project, San Isidro. Some of the data were recorded by the author in several government reports and published in the Proceedings of the Second International Conference on the Transfer of Water Resources Knowledge.[1] The source of data is primarily field work which extended over a period of several years in the aforementioned areas. It includes systematic observations and interviews pertaining to how resource managers use their water and land resources.

These were supplemented over a period of four years by observations which the author made while employed with the Peruvian National Office for Agrarian Reform (ONRA) and for the Peruvian National Office for the Evaluation of Natural Resources (ONERN). Thus, the interpretation of the field studies is seasoned by the later experience of talking and working with irrigation farmers in Arequipa and other areas along the Peruvian arid coast.

The Data

Specifically, information was obtained on scheduling irrigation

---

[1] R.C. Holmes, "Evaluación de Riegos - Consumos de Agua." Paper presented at the III Congreso Nacional de Ingenieros Agrónomos, Lima, Perú, 19-25 September 1965; ________, "La Evaluación del Uso del Agua de Regadío en la Campiña de Arequipa y la Irrigación, La Joya y su Ampliación." A report to the Oficina Nacional de Reforma Agraria del Perú, Arequipa Office, March 1968 (typewritten); ________, "The Transfer of Irrigation Water Knowledge in the Developing World: Irrigation Water Use in the Arequipa Region of Southern Peru," in Water Knowledge Transfer, ed. Neil S. Grigg, Proceedings of the Second International Conference on Transfer of Water Resources Knowledge, June 1977, Fort Collins, Colorado, vol. 2 (Fort Collins, Colorado: Water Resources Publications, 1978), pp. 977-984.

waterings for three farms close to the city of Arequipa and 14 farms in the original La Joya Project along with some general information pertaining to the extended area of the La Joya Project. The amount of water applied per irrigation was determined for 20 fields near the city of Arequipa and 18 fields in the La Joya Project. Also, soil analysis, including soil moisture conditions before waterings, were determined for 12 fields located near the city of Arequipa and on the La Joya Project and its extension.

On all the farms, detailed field maps were prepared, indicating the field layout along with the water delivery system. In addition, these field maps show the physical characteristics of the irrigation methods employed.

Land use information was collected for an entire year on a field-by-field basis under the command of the ditch in the intensively studied farms near the city of Arequipa, and in the 14 farms in the La Joya Project.

While the above information was being gathered, non-structured interviews were conducted among five or more farmers near the city of Arequipa, 13 farmers in the La Joya Project, and six farmers in the extended area of the La Joya Project. Interviews or discussions also were held with at least a dozen individuals of the local technical-scientific community, as well as with several of national and international rank. The failure of farmers to accept a new project was documented from the La Joya records.

In reporting the findings, the human geography of the area is examined in chapter II. Chapter III discusses the physical environment as modified by irrigated agriculture. The principal results of the field work are given in chapter IV, entitled "Irrigation Water Control in Arequipa." The major findings are presented in chapter V. In these, lessons are drawn which may be relevant in future irrigated agricultural development programs.

CHAPTER II

THE HUMAN GEOGRAPHY OF IRRIGATED AGRICULTURE IN AREQUIPA

The Quilca Basin has a long history of irrigated agriculture that predates even the Inca civilization once found in the region. The general settlement pattern has been one of locating sites off the limited arable land, in villages. In recent times, on the newer irrigation schemes, homesteads have been located on individual farms resulting in a dispersed settlement pattern. Many farmers have reacted negatively to this causing planners to rethink this issue for future projects.

Population growth, brought about by natural increase as well as immigration from the Lake Titicaca Basin, has created increased pressures on the limited resource base, especially irrigated land. As a result there is interest in increasing the arable area.

Only a small part of the Department of Arequipa is in farms, which are mainly used for growing truck crops and raising cattle to support the important milk industry. These intensively cultivated areas are dependent on irrigation, but, even so, only a relatively small area is irrigated. Many small-to-medium-sized irrigated farms cluster upon the landscape. Most of them are either family/owner-operated or rented. Farmers in the Chili Basin base their agricultural practices largely on practical experience. Since farming is the principal source of work, there is always a steady flow of young people entering the agricultural field. Many of the farms, particularly in the older sections, are not in a single block but consist of small parcels of land separated from each other. Farm produce reaches Arequipa, Lima, or international markets by way of a middleman or the farmer himself. Also the area is well served by a system of roads and railroads.

Knowledge of the attitudes and perceptions of the various groups in the study area aids in understanding the management of irrigated lands. The groups singled out for brief comment on their behavioral patterns are the Indian, the _Cholo_, the Middle Class, and the Upper Class.

This general description of the economic conditions is amplified with data from various government reports, among which the agricultural census of Peru for 1961 is of fundamental importance, and which is supplemented by data from the mid-1950s and the 1970s.

Early History and Settlement Pattern

The Quilca River Basin in the Far South of Peru was settled before the coming of the Spaniards by both Inca and pre-Inca peoples. The first great pre-Incan cultural group to be found in the Far South of Peru, the Tiahuanaco Empire, lasted for 200 years (600 A.D.-800 A.D.). Apart from conquering the Titicaca Basin, this group extended its control into coastal and highland northern Chile as well as the Quilca Basin. This was the most culturally advanced pre-Inca group to rule over the Far South. The earlier, and the first "great style" of ancient art and architecture of Peru, the Chavín Culture, had spread along the entire coast with the exception of the Far South.[1] During the period of Tiahuanaco Empire, the interchange of goods and ideas across the Central Andes increased greatly. Also during the pre-Inca times, irrigated agricultural terraces were constructed in the general vicinity of the study area of the Chili Basin. After the fall of the Tiahuanaco Empire, 600 years elapsed until the next great culture appeared in the 1400s. This was the Inca Empire. In 1471 Topa Inca extended the Inca Empire nearly to its ultimate limits, including the Far South of Peru.[2] In the early 1500s the Inca Empire was subdued by the Conquistadores, and in 1540 Don Garcí Manuel de Carvajal founded the Spanish city of Arequipa.

Pre-Columbian people in the Quilca Basin, as well as along adjacent coastal valleys, had at one time or another engaged in hunting, food gathering, fishing, floodwater farming, and irrigated agriculture. These groups established settlements at different sites the location being conditioned by the economy of the area. Where agricultural land was at a premium, settlements were located on land, not needed for farming. The situation described by Farrington for the Moche Valley on the north coast of Peru is an apt description of the early conditions in the Quilca Basin. "Sites were often above and adjacent to the uppermost canal or on low mounds and hillsides

---

[1] Edward P. Lanning, Peru Before the Incas (Englewood Cliffs, New Jersey: Prentice-Hall, Inc., 1967), pp. 96-97.

[2] Ibid., p. 158.

within the cultivated area."[1] Examples of settlements in the Chili Basin located off agricultural land are the pre-Inca settlement at Churajón, about the terraced, irrigated, (but now abandoned) farm land, and the Inca-founded hilltop settlements of Chiguata, Paucarpata, Characato, Socabaya, Tiabaya, and Yanahuara, which are located in the surrounding countryside or Campiña of Arequipa (Campiña).[2]

There was also a tendency in this basin for people to settle in villages instead of living on isolated farms scattered over the land. Even after the fall of the Tiahuanaco Empire, when all the cities of the southern highlands were abandoned and inhabitants scattered over the countryside, the population according to Lanning "seems to have lived in villages and small rural towns. As for the Inca culture, notwithstanding the fact that their settlement system was rural in concept, their policy was also the setting of their subjects in villages and small towns."[3]

A similar settlement pattern continued under the Spanish, who occupied existing villages or created new ones.[4] Many of these settlements were located in valleys, often on useless, sterile lands but near a good supply of surface water that could be used for irrigation. In certain areas, where they settled in existing villages, as in the case of Arequipa, the Spanish extended the original irrigation system in order to accommodate a larger population.

An important aspect of life during pre-Colonial times was the development of a communal form of living. However, since irrigated agriculture was an important part of the life of these people, the chief (cacique) of the village controlled the operation and maintenance of the village irrigation system. The village form of settlement, together with irrigated agriculture, set the stage for this complex sociopolitical organization. Those engaged in agriculture went out each day to work their nearby farms either alone or cooperatively.

Those villages that lacked sufficient precipitation for cropping established their own irrigation systems, some of which are still in operation today, as is the extensive system in the vicinity

---

[1] Ian Farrington, "Irrigation and Settlement Patterns, Preliminary Research Results from the North Coast of Peru," in Irrigation's Impact on Society, ed. Theodore E. Downing and McGuire Gibson, Anthropological Papers of the University of Arizona, no. 25 (Tucson: The University of Arizona Press, 1974), p. 84.

[2] Victor Benavente, Historia Sintética de Arequipa (n.p., 1940) p. 15.

[3] Lanning, Peru Before the Incas, p. 163.

[4] Julian H. Steward and Louis C. Faron, Native Peoples of South America (New York: McGraw-Hill Book Company, Inc., 1959), p. 156.

of the city of Arequipa. In the Campiña several of the irrigation canals still functioning were built by the Incas. These include the Antiquilla, Acequia Alta, and San Jerónimo Canals.[1]

In recent times, with the development of new irrigation projects, farmers were settled on their farms rather than in villages, as was customary in the past. As long as this occurred near the urbanized area of Arequipa, as exemplified by the irrigation project of Zamácola, no great adverse effects were manifested. However, this drastic change in settlement pattern has taken place also at some distance from the city of Arequipa in the new irrigation projects on the Pampa of La Joya. Difficulties encountered by farmers in adjusting to this form of farm layout, which isolates them from their neighbors and communal activities, has led to an impasse between the farmer and the agricultural planner. In devising the settlement plan in these irrigation projects, planners did not consider the effect of this new settlement pattern on the life of the farmers. In many instances farmers have been reluctant to leave an established community, with adequate schools, shops, and churches to live in an isolated non-village type of settlement having only makeshift housing. This condition, which exists in the original La Joya irrigation project, may account in part for the relatively large owner absenteeism found there. The same condition prevailed in the early stage of colonizing the first major extension of the La Joya Project (San Isidro). Here the situation became acute due to the lack of readily available loans on the international market for providing decent housing. Certainly one of the most pathetic sights the author saw in this new project was a fire that consumed a reed-constructed home belonging to one of the new farmers. Within a matter of minutes the farmer lost all of his meager possessions. A malfunctioning kerosene stove caused this disaster.

Recognizing that the ingrained effect of village communal living could not be overcome, planners are once again locating farm homes in small clusters in the newer extensions of the La Joya Project.[2] It is hoped that this will provide greater incentive for the new farmers to reside on these irrigated projects, instead of moving back to the city and letting overseers (mayordomos) operate their farms.

---

[1]Alberto de Rivero, Arequipa en su IV Centenario (n.p., 1940), p. 85.

[2]Otto Schultz Wertzel, et al., Proyecto de Planificación, Colonización "La Joya." Programa Conjunto OEA-Israel, Equipo Agropecuario Peruano Israeli (Arequipa, Perú: Oficina Nacional de Reforma Agraria, 1967).

Population Growth

As a result of early colonial policy, the native population was sharply reduced and, for the country as a whole, it did not again reach preconquest levels until the 1940s.[1] Along with the present rapid growth rate of around 3 percent per annum for Peru, there is a very large sector of the population below 15 years of age to be considered. This preproductive group may bring about an even faster growth in population than is presently being experienced.[2]

In addition to this rapid natural increase, which exists nationwide, there also has been a large movement of Indians from the Sierra around Lake Titicaca into the Chili Basin. Many conditions in the Lake Titicaca area account for this mass migration. This area is one of the most densely settled areas in all of South Peru,[3] and many Indians have migrated, of necessity, because of the growing population pressure on the resource base of this lake basin. The innumerable natural hazards that occur periodically also have led to mass migration of Indians from the Lake Titicaca Basin from time to time. The 1950s was a period of great stress for the residents of this area. Dew noted that, in the weather crisis which occurred during the 1959-60 period:

> The rains during the agricultural season were exceedingly heavy. On February 2, 1960, it was reported that 17,000 hectares in the Province of Huancané had been flooded by the Ramis River affecting the crops and homes of over 15,000 peasants. . . . Then in the middle of March, a premature end of rain was followed by a series of disastrous frosts, which destroyed an estimed 90 percent of the crops in the Province of Chucuito, according to the SCIPA agent in Puno. . . . It was estimated that the damage in Puno caused by the "black frost" might be worse than that which the area suffered in 1956-57. Indians were reported to be pouring into Puno for help and proceeding on to Juliaca and Arequipa. Children were again reported being sold by their parents.[4]

Thus, one of the principal destinations for people fleeing the Lake Titicaca area is the Chili Basin, with its large urban center of Arequipa. True, some of the refugees go on to Lima, but many stay in the local area to create the shantytowns (pueblos jóvenes) that ring the city of Arequipa. With a limited industrial base,

---

[1] Richard P. Schaedel, La Demografía y Los Recursos Humanos del Sur del Perú, Serie: Antropología Social, no. 8 (Mexico, D.F.: Instituto Indigenista Interamericano, 1967), p. 7.

[2] Ibid., p. 12.

[3] Ibid., p. 2.

[4] Edward Dew, Politics in the Altiplano: The Dynamics of Change in Rural Peru (Austin: University of Texas Press, 1969), pp. 112-113.

as well as limited arable lands, Arequipa is hard-pressed to absorb these new residents. One way to meet the growing population pressure in the Chili Basin is continued agitation for more irrigation projects, as exemplified by the recent irrigation activities on the Pampa of La Joya, and the continuing interest in irrigating the adjacent Pampa of Siguas by harnessing the Majes River, the major river to the west of the Quilca Basin.[1]

Land in Farms

Agriculture in the province of Arequipa is more intensive than in the Department of Arequipa as a whole (table 1). Although 11 percent of the total area of the Department is in farms, compared to only a little over 3 percent for the Province, 63 percent of the farm land in the Province is cultivated, compared with a low of 10 percent for the Department (table 1). Furthermore the Province has a slightly greater percentage of farm land under irrigation (78 percent) than the Department (71 percent).

The three areas studied in the Province vary in total area, with the Campiña having almost 9,200 hectares in farms, La Joya Antigua almost 3,500 hectares, and San Isidro 1,000 hectares (table 1). Of the three areas a greater percentage of farm land in the Campiña is irrigated regularly. In La Joya Antigua it is not uncommon for one-fifth of the farms to be fallow for several years. Since San Isidro had just begun to be settled at the time of this study, it was not surprising to find that only one-third of a farm was being cultivated, the remaining portion of the holding being not yet prepared to receive the new water supply.

Land in Crops

Unlike other Departments along the Pacific coast of Peru, especially those along the central and northern coasts, the Department of Arequipa concentrates on the production of food and forage crops rather than industrial crops (table 2). The principal food crop is cereals, of which wheat and corn are the most widely planted (table 3). A significant amount of Irish potatoes and green vegetables is also grown. The chief forage crop planted on improved pasture land is alfalfa, and the principal industrial crops in the Department are cotton and sugar cane.

The mix of crops grown in the Province of Arequipa is similar

---

[1]Rosendo F. Chávez Díaz, El Problema de Majes y Siguas: Planificación Regional de Irrigación en Arequipa (n.p., 1959).

TABLE 1

USE OF LAND IN DEPARTMENT OF AREQUIPA, 1961

	Department of Arequipa	Province of Arequipa	Campiña Arequipa	La Joya Antigua Irrigation	San Isidro Irrigation
Total Area, Ha	6,352,700	837,700	-----	-----	-----
Total Land Farms, Ha	698,475	26,209	9,177	3,433	1,000
Percent of Total Area in Farms	11.0	3.1	-----	-----	-----
Area of Farms Cultivated, Ha	71,083	16,482	9,177	2,650	333
Percent of Total Land in Farms Cultivated	10.2	62.9	100	77.2	33.3
Irrigated Farm Land, Ha	50,859	12,845	9,177	2,650	333
Percent of Cultivated Farm Land Irrigated	71.5	77.9	100	100	100

SOURCES: República del Perú, Ministerio de Hacienda y Comercio. Dirección Nacional de Estadística y Censos. Censos Nacionales de Población, Vivienda y Agropecuario: 1961 Departamento de Arequipa (Censo, 1961) (Lima, 1969), p. 217, and records of ONERN and the Junta de Desarrollo de Arequipa.

TABLE 2

LAND USE BY CROPS, DEPARTMENT OF AREQUIPA, 1955

Type Crops	Thousands of Hectares	Percent
Food	51	62
Forage	23	28
Industrial	8	10
Total	82	100

SOURCE: República del Perú, Ministerio de Fomento y Obras Publicas, Evaluación Económica y Planeamiento Departamento de Arequipa (SCIF, 1958) (Lima: Prensa y Publicaciones, SCIF, 1958), p. 34.

to the Department as a whole, with a few exceptions (table 3). Since the Province is located at a somewhat high elevation, between 1,500 m to 2,500 m, very little cotton and no sugar cane are grown. The major emphasis here is on forage crops, with 43 percent of the land in alfalfa. Cereals are second with 36 percent, of which somewhat over half is in wheat. The bulk of the remainder of the cereals consists of corn. Percentage-wise, the amount of vegetables grown is about the same as it is for the Department; however, the composition of the mix is different. A greater percentage of the land in the Province is planted in fruit trees, garlic, and onions than in the Department as a whole.

The Campiña, La Joya Antigua, and San Isidro vary as to agricultural plantings. The greatest crop intensification is found in the Campiña area. Since the land in new irrigation projects is customarily sown in alfalfa to condition the soils, the San Isidro farmers initially planted all the land in alfalfa along with some double croppings of barley. The types of crops grown in the Campiña and in La Joya Antigua are practically the same but amounts do vary (table 4). Because there have not been any dramatic shifts in the land-use pattern between the Campiña and the La Joya Antigua areas, it is felt that a ten-year span in data between the two does not invalidate the general observations made. Both areas greatly emphasize alfalfa, La Joya Antigua showing a concentration of approximately 80 percent. Of the remaining hectarage in this area, about 15 percent is planted in corn, potatoes, and wheat, with the remaining 5 percent largely in onions, garlic, tomatoes, and fruit trees

TABLE 3

LAND USE BY CROPS, DEPARTMENT AND PROVINCE OF AREQUIPA, 1955

Crops	Department of Arequipa				Province of Arequipa			
	Thousands of Hectares (Group)	Thousands of Hectares (Item)	Percent (Group)	Percent (Item)	Thousands of Hectares (Group)	Thousands of Hectares (Item)	Percent (Group)	Percent (Item)
Tuberous	6.4		9.0		1.3		7.7	
Potatoes		6.4		9.0		1.3		7.7
Cereals	24.4		34.3		6.1		36.3	
Wheat		7.4		10.4		3.7		22.0
Corn		9.4		13.2		2.1		12.5
Barley		4.0		5.6		0.3		1.8
Rice		2.8		3.9				
Vegetables and Fruits	6.9		9.7		2.1		12.5	
Green Vegetables		4.3		6.0		1.3		7.7
Beans (Habas)		0.1		0.1		0.1		0.6
Sweet Potatoes		0.4		0.6		---		---
Yucca		0.1		0.1		---		---
Fruit		2.0		0.3		0.7		4.2
Tropical	13.3		18.7		0.1		0.6	
Cotton		10.3		14.5		0.1		0.6
Sugar Cane		2.6		3.7		---		---
Other (excluding Coffee and Cocoa)		0.4		0.6		---		---
Cultivated Pastures	20.1		28.3		7.2		42.9	
Alfalfa		20.1		28.3		7.2		42.9
Total	71.1		100		16.8		100	

SOURCE: Plan Regional para el Desarrollo del Sur del Perú, Manual de Estadística Regional (Lima: Servicio Cooperativo Peruano Norteamericano de Educación, 1959) p. 139-146.

TABLE 4

LAND USE BY CROPS, CAMPIÑA, AND LA JOYA ANTIGUA 1963 AND 1972

	Campiña, 1972		La Joya Antigua, 1963	
Crops	Ha	Percent	Ha	Percent
Alfalfa	5,331	62	2,280	82
Wheat	103	1	80	3
Corn	862	10	150	5
Barley	92	1	5	--
Potatoes	220	3	150	5
Onions	1,202	14	30	1
Garlic	278	3	--	--
Tomatoes	24	--	50	2
Fruits	88	1	15	1
Other	394	5	30	1
Total	8,594	100	2,790	100

SOURCES: Compiled from records of ONERN and the Junta de Desarrollo de Arequipa.

(table 4). In contrast, the Campiña devotes 60 percent of the cropland to alfalfa, 14 percent to onions, 10 percent to corn, 3 percent each to garlic and potatoes, and the remaining land to wheat, barley, and additional fruits and vegetables.

The value of crop production varies within the Department of Arequipa, reflecting in part the wide range of crops grown in different climatic zones. Care should be taken in utilizing the data pertaining to the value of production per hectare reported in this study since the information for the Department and Province of Arequipa as a whole represents the situation in 1957, while the data for the three study areas within the Province reflects the conditions of 1971-72. Thus, this data will be used only in comparing the situation within, and not between those years.

The average value of crop production for the Province of Arequipa was about S/. 4,500 per hectare, which is somewhat higher than that for the Department of Arequipa, with S/. 4,000 per hectare (tables 5 and 6). The spread of value of production within the areas is indicated by the two groupings of crops above and the two groupings of crops below the average. For the Department as a whole, 62 percent of the land produces crops with a value of production below the average; this includes such crops as sugar cane, cereals, and alfalfa. The remaining 38 percent of the land has crops above the average in value of production; exotic tropical crops being well above the average with S/.36,000 per hectare. Other high value crops include fruits, potatoes, cotton, vegetables, and rice. The Province is somewhat similar to the Department in that the majority (80 percent) of the land is in crops that fall below the average, cereals and alfalfa. The remaining one-fifth of the land above the average is planted in potatoes, fruits, vegetables, and cotton (table 6).

On the whole, the value of production in the Campiña is far superior to that found in La Joya Antigua (tables 7 and 8). The average value of production of S/.41,000 per hectare for the Campiña is 155 percent greater than the value of production for La Joya Antigua.

In the Campiña, 75 percent of the land has a value of production below the average (table 7). All of the forage crops and cereals, as well as some of the vegetables, are found in this lower-than-average group. The remaining 25 percent of the land is far above the average, ranging from almost S/.50,000 to S/.137,000 per hectare. This latter figure corresponds to the production of onions. Additional vegetables in this higher-than-average category are garlic, scallions, and potatoes. About 85 percent of the land of La Joya Antigua has a value of production below the average of S/.26,540 per hectare (table 8). As in the Campiña, the crops in this category are alfalfa, grains, beans, and fruits. In the above-average group, which includes the remaining 15 percent of the land, garlic, onions, tomatoes, additional vegetables and vineyards are recorded. Since San Isidro was so new at the time of this study, no meaningful information was available concerning value of production.

Number and Size of Farms

A little over 20 percent of the 24,540 farms in the Department of Arequipa are located in the Province of Arequipa, while only 4 percent of the 698,475 hectares in farms in the Department are found in the Province (tables 9 and 10). Over 85 percent of the farms in the Province are found in the Campiña but only about

TABLE 5

VALUE OF PRODUCTION PER HECTARE, DEPARTMENT OF AREQUIPA, 1957

Groups	Value of Production per Hectare	Major Crops	Percent of All Irrigated Land
I	S/.9,250 to S/.36,000	Tropical Crops (excluding Coffee)	0.6
II	S/.5,200 to S/. 9,249	Fruits, Vegetables, Cotton, Rice	37.0
III	S/.2,574 to S/. 3,308	Sugar Cane, Wheat, Corn	27.3
IV	S/.1,550 to S/. 2,100	Alfalfa, Barley (Cereal)	35.0

SOURCE: See table 3.

TABLE 6

VALUE OF PRODUCTION PER HECTARE, PROVINCE OF AREQUIPA, 1957

Groups	Value of Production per Hectare	Major Crops	Percent of All Irrigated Land
I	Greater than S/.13,000	Potatoes	7.8
II	S/.6,000 to S/. 9,000	Fruits, Vegetables, Cotton	12.6
III	S/.3,857 to S/. 4,243	Wheat, Corn	34.7
IV	S/.2,097 to S/. 2,333	Barley, Alfalfa	44.9

SOURCE: See table 3.

TABLE 7

VALUE OF PRODUCTION PER HECTARE, <u>CAMPIÑA</u>, 1971-1972

Groups	Value of Production per Hectare	Major Crops	Percent of All Irrigated Land
I	S/.84,000 to S/.136,800	Onions, Garlic	21.5
II	S/.49,000	Potatoes	3.4
III	S/.20,400 to S/. 28,000	Other Vegetables, Fruits, Alfalfa	60.0
IV	less than S/. 16,000	Cereals, other Forage Crops	14.6

SOURCE: Compiled from records of ONERN.

TABLE 8

VALUE OF PRODUCTION PER HECTARE, LA JOYA ANTIGUA IRRIGATION, 1971-1972

Groups	Value of Production per Hectare	Major Crops	Percent of All Irrigated Land
I	S/.52,500 to S/.117,600	Garlic, Onions, Tomatoes	11.2
II	S/.28,180 to S/. 38,700	Other Vegetables, Vineyards	3.5
III	S/.19,200 to S/. 23,400	Fruits, Beans (Habas), Alfalfa	68.8
IV	S/. 2,400 to S/. 14,350	Corn (Forage) Wheat, Barley, Forage Crops	16.5

SOURCE: See table 7.

TABLE 9

NUMBER AND SIZE OF FARMS, DEPARTMENT OF AREQUIPA, 1961

Size of Farm Ha	Farms		Area	
	Number	Percent	Ha	Percent
Less than 1	11,304	43.3	5,123.3	0.7
1 to 5	9,728	37.3	20,712.0	3.0
5 to 500	3,186	12.2	194,684.5	27.9
500 to 1,000	214	0.8	142,489.8	20.4
1,000 to 2,500	86	0.3	110,794.7	15.9
2,500 to 5,000	12	0.1	44,838.6	6.4
5,000 to 10,000	7	---	48,017.2	6.9
over 10,000	3	---	131,705.2	18.8
Total	24,540	100	698,475.3	100

SOURCE: Censo, 1961, pp. 218-219.

one-third of the provincial farm land is located in this area (tables 10 and 11). Somewhat over 300 farms, which represent only a small proportion of the farms in the Province, are located in La Joya Antigua (table 12). Likewise, only a small amount of the farm land of the Province, 3,500 hectares, is found in this irrigation project. The initial stage of settlement of San Isidro consists of 100 farms, with a total of 1,000 hectares.

The average size of farms in the Province is about 5 hectares, which is about one-sixth the average size of farms for the Department, 28 hectares (tables 9 and 10). In both the Department and Province approximately 80 percent of the farms are under 5 hectares. There are about 100 farms in the Department over 1,000 hectares, but in the Province only two farms are of this magnitude.

In terms of the size of the farms, there is a great contrast between the Campiña and the other two study areas. The Campiña has many extremely small farms, the average size is 2 hectares. The existence of these extremely small farms is considered a negative farming condition known as minifundismo. What is more, over 90 percent of the farms in this region contain less than 5 hectares,

TABLE 10

NUMBER AND SIZE OF FARMS, PROVINCE OF AREQUIPA, 1961

Size of Farm Ha	Farms		Area	
	Number	Percent	Ha	Percent
Less than 1	2,854	48.4	1,173.0	4.5
1 to 5	1,866	31.7	4,167.5	15.9
5 to 500	757	12.8	13,723.4	52.4
500 to 1,000	---	----	---	----
1,000 to 2,500	---	----	---	----
2,500 to 5,000	2	----	7,144.6	27.2
5,000 to 10,000	---	----	---	----
over 10,000	---	----	---	----
Total	5,479	100	26,208.5	100

SOURCE; Censo, 1961, pp. 220-221.

and there are none greater than 100 hectares. On the other hand, the average size of farms in La Joya Antigua is about 10 hectares. While only one-third of the farms in this area are under 5 hectares, only 2 farms are over 100 hectares. The initial colonization in San Isidro allocated 10 hectares per farmer. The original intent had been to establish farms of 25 hectares, but because of the desire to settle more farmers, the size was reduced.

Farm Tenure and Fragmentation

While owner-operated farms are relatively common in both the Department and the Province of Arequipa as a whole, this is less true in the Campiña and especially so in La Joya Antigua (tables 13 and 14).

Before the enactment of the new agrarian reform program in 1969, there had been a tendency for owner-operated farm land to decrease with time, as more and more of the landowners rented all, or part, of their holdings. Now this trend is being reversed.

Nevertheless, it is believed that land tenure had not changed dramatically in the period 1955 to 1961 and so the following general

TABLE 11

NUMBER AND SIZE OF FARMS, CAMPIÑA AREQUIPA, 1955

Size of Farm Ha	Farms		Area	
	Number	Percent	Ha	Percent
Less than 5	4,378	92.4	5,291	57.7
6 - 10	263	5.6	1,920	20.9
11 - 15	54	1.2	674	7.3
16 - 20	16	0.3	328	3.6
21 - 30	7	0.1	186	2.0
31 - 50	12	0.3	437	4.8
51 - 100	4	0.1	341	3.7
over 100	0	0	0	0
Total	4,684	100	9,177	100

SOURCE: SCIF, 1958, Anexo. 11.

TABLE 12

NUMBER AND SIZE OF FARMS
LA JOYA ANTIGUA IRRIGATION, 1955

Size of Farm Ha	Farms		Area	
	Number	Percent	Ha	Percent
less than 5	115	35.7	576	16.8
6 - 10	135	41.9	998	29.1
11 - 15	30	9.3	370	10.8
16 - 20	15	4.7	283	8.2
21 - 30	14	4.4	408	11.9
31 - 50	9	2.8	278	8.1
51 - 100	2	0.6	200	5.8
over 100	2	0.6	320	9.3
Total	322	100	3,433	100

SOURCE: See table 11.

TABLE 13

LAND TENURE, DEPARTMENT AND PROVINCE OF AREQUIPA, 1961

Type of Tenure	Department of Arequipa				Province of Arequipa			
	Farms		Area		Farms		Area	
	Number	Percent	Ha	Percent	Number	Percent	Ha	Percent
Owner-operated	15,614	71.2	524,117	77.2	2,938	60.6	16,888	70.0
Rented	4,403	20.0	72,986	10.8	1,743	36.0	6,817	28.3
Share-crop	1,568	7.2	50,963	7.5	94	2.0	141	0.7
Indian Community	13	0.1	5,832	0.9	0	0	0	0
Other	325	1.5	24,933	3.7	70	1.4	276	1.1
Total	21,923	100	678,831	100	4,845	100	24,122	100

SOURCE: See table 9.

TABLE 14

LAND TENURE CAMPIÑA AREQUIPA AND LA JOYA ANTIGUA, 1955

Tenure Classes and Size Parcels (Ha)	Campiña Arequipa		La Joya Antigua	
	Area Ha	Percent	Area Ha	Percent
Owner-operator				
0 - 5	3,004	33	180	5
6 - 10	532	6	492	14
10 +	1,140	12	676	20
Subtotal	4,676	51	1,348	39
Renters				
0 - 5	2,287	25	396	12
6 - 10	1,388	15	506	15
10 +	826	9	1,183	34
Subtotal	4,501	49	2,085	61
Total	9,177	100	3,433	100

SOURCE: SCIF, 1958, Anexo 10.

observations seem valid. In both the Department and Province of Arequipa, over 60 percent of the farms are owner-operated, with a somewhat greater percentage of the farm land being operated by owners in the two areas (table 13). The bulk of the remaining farms and farm land is rented; only a very small fraction is under other forms of tenure, such as sharecropping or communal Indian lands.

Farm tenure in the Chili Basin varies greatly. While all the farm land in San Isidro is owner-operated, only a little over 50 percent is owner-operated in the Campiña and not quite 40 percent in La Joya Antigua (table 14).

Another characteristic of agriculture in the Province of Arequipa is the fragmentation of farms. This is especially prevalent in the Campiña, exists in a more limited manner in La Joya Antigua, and does not exist at all in San Isidro. In the Campiña, fragmentation consists of a scattering of small parcels, while in La Joya Antigua entire tracts of land, consisting of several parcels, are scattered throughout the project.

These parcels in La Joya Antigua, in reality former farms originally laid out during the initial colonization of the area, consist of 5 or more hectares. Here a farmer may have, in addition to his own five-or-ten-hectare farm, additional rental farms in the general vicinity. In this way he can raise sufficient alfalfa for his herd of milk and/or beef cattle and not be forced to buy alfalfa from his neighbors to feed his cattle during the winter.

Farmers

Since farming is the principal livelihood of the people of Arequipa, farmers mirror the general demographic and social characteristics of the group as a whole. The age of farmers is not appreciably different from that of the general population, although many of the original farmers who colonized La Joya Antigua about 35 years ago are still managing farms and have yet to pass their holdings on to their sons. With this condition, it is possible that there may be a tendency of slower adoption of innovations in La Joya Antigua than in the Campiña. Perhaps this, in part, explains why the La Joya Antigua Irrigation Project has never, since the early days of settlement in 1940, made substantial improvements in the management of land and water resources beyond those reflected in the regional milieu.

Most farmers in the Chili Basin are familiar with farming procedures, as are their sons who inherit the farms. There are, however, situations where the farming population is not always acquainted with the most appropriate procedures to follow. A renter acquiring a new piece of land (even though he has been in farming for years) may at first be unfamiliar with the new environment, especially the most effective manner of applying irrigation water. But if this newly acquired land is near an area the renter has been working for some time, uncertainty of management practices may not be a major issue. Also if the renter hires a farm laborer who has worked on farms in the general area, and who is thus familiar with the land and water resources, continuity in management practices will continue.

With the colonization of San Isidro by the government in 1967, the new farmers confronted an entirely new situation which put on trial their accustomed farm practices. Even the most experienced farmers need a period of adjustment to work out a proper farm management program; the first 3 or 4 years being comparable to the shakedown time of a new industrial plant. In San Isidro, the government gave the farmers a grace period of 2 years in which to have their farms producing. By the third year the farmers were expected to begin payment on their farms.

In order to qualify for a farm in the San Isidro project, a person had to be actively engaged in farming, as a farm owner, renter, or farm laborer. Some would naturally have more experience than others. Former farm laborers, with limited involvement in farm decision-making, might be more willing to accept new irrigation practices than those with a more intimate acquaintance with farming.

The sources of information used in managing farming enterprises in the area are influenced, to a degree, by outside aid, either through agricultural extension agents or literature. Other than what appears in the local newspaper, few have a chance to read about new farming innovations. Some farmers receive information from extension agents on the use of better seeds, on plant varieties, and how to combat insects and plant diseases. With the exception of San Isidro, farmers receive little technical aid regarding the best way to manage irrigation water. The effort of technicians to help farmers manage their water supply more judiciously in San Isidro resulted in mixed benefits. This episode will be elaborated upon in a later chapter.

Marketing and Transportation

The marketing of agricultural produce of the Chili Basin is controlled by various entities. Whereas some farmers sell their produce directly to a market, as in the city of Arequipa, the usual practice is to sell to a middleman (intermediario), who buys the farm produce in the field and carries it to the various markets.

There are two major exceptions in this marketing procedure. All milk destined for the Leche Gloria processing and canning plant, located in the city of Arequipa, is marketed by the company itself. The company sets the price of milk, based on the butter-fat content, and collects milk daily from each farmer who has a contract with the company. Another manner of marketing farm produce is through a farm cooperative. This procedure is followed by the farmers in San Isidro in selling barley and excess alfalfa.

The Chili Basin is well served by roads and railways. The dense network of farmroads gives most farmers easy access to market. With the completion of a new paved highway linking the city of Arequipa with the Pampa of La Joya, truckers now have gentler grades to negotiate when traveling to this major market center of the South. Farmers located on the Pampa of La Joya are only about 1 hour from the city of Arequipa. The Pan American Highway, which bisects the

Chili Basin, facilitates the trucking of farm produce to Lima, the main national market of the country. Shippers of garlic, destined for overseas markets, have ready access to the Pacific port of Mollendo/Matarani either by road or railway. Likewise, farmers of the Chili Basin have access to the markets in the Departments of Puno and Cuzco, by either road or railway.

Perceptions and Attitudes of the People

Knowledge of the behavior of a people assists in unraveling the mysteries surrounding the actual operation of irrigated agriculture and leads to a better understanding of how social change takes place in such a special environment. For Boulding and others the most important determinant in human behavior is the individual's and society's perception of the world, and the working of inductive reasoning as reflected in the folk images that are held by ordinary people as a result of their ordinary experiences.[1] Thus, much of our behavior results as Boulding says from:

> . . . the process of generalization or induction from [one's] personal experience. [The individual] projects what he has experienced in the past into the future, and he lives wholly in a world of what William Blake calls "minute particulars."[2]

Therefore, there is a great need to evaluate public attitudes and perceptions as they impinge upon the management of resources. As Biswas and Durie stated, "Most studies indicate that the disregard of people's desires and wishes have generally resulted in the failure of water resource projects in achieving their desired objectives."[3] An example of human behavior hindering the effective management of irrigation water is related by Cantor in telling of the superstitious barriers to using irrigation water in the Hirakud project in India.

> [in this area] one of the prevailing beliefs is that one cannot grow rice or any other crop on the same land twice in the same year, even though the water provided by the scheme now makes this practicable. There is a further local belief that once water has been used for the generation of power, it loses some vital 'electricity' that is essential to promote plant growth. As a result, many peasants in this area

---

[1]Kenneth E. Boulding, "Social Sciences," in The Great Ideas Today, eds. Robert M. Hutchins and Mortimer J. Adler (Chicago: Encyclopedia Britannica, 1965), p. 256.

[2]Ibid.

[3]Asit K. Biswas and Robert W. Durie, "Sociological Aspects of Water Development," Water Resources Bulletin 7 (December 1971):1142.

do not readily agree to use the water even when it is demonstrated that by doing so they can rapidly double their yields and their incomes. In an effort to overcome this prejudice, the state government is not only giving irrigation water free for the time being, but is also constructing the field channels as well, a task that is normally the responsibility of the farmer.[1]

The values held by Peruvians can be briefly reviewed in light of folk images that have relevance to economic development in general and irrigated agriculture in particular. Even though much of the information is taken from the literature, the author's residential intimacy with these groups permits him an awareness and sensitivity to problems and frustrations which they often encounter. In Peru, the various subcultures of society cannot be clearly delineated, in that no group has developed in isolation. Among the two most startlingly different groups, the Indian and the Upper Class White, there has been a sharing of values. Many whites accept native superstitions and they are not immune to Indian influence in foods, material features, and folk belief.[2] Several students of the Peruvian Indian have noted that, through their innumerable contacts with non-Indian groups, the truly native peoples have absorbed much of the outside values of the world, yet have retained some of their own unique values.[3] So this intermingling of groups makes it difficult to effectively analyze one form of social organization in isolation from the other.[4] Nevertheless, in order to gain a cursory but somewhat systematic understanding of how each subculture affects the environment and, in turn, is affected by the environment, the four major groups - the Indian, the Cholo, the Middle Class, and the Upper Class - will be briefly described.

Indians

In southern Peru, about 50 percent of the population consists of various groups of Indians, chief among these are the Quechuas and Aymaras, who live almost entirely in the rural area. In the Department of Arequipa, only about 2 percent of the inhabitants are classed as Indian. However, the Department on its eastern

---

[1] Leonard M. Cantor, A World Geography of Irrigation (London: Oliver and Boyd, 1967), p. 90.

[2] Steward and Faron, Native Peoples of South America, pp. 153-154.

[3] Schaedel, La Demografía y Los Recursos Humanos del Sur del Perú, pp. 23-24.

[4] Eric R. Wolf, Sons of the Shaking Earth (Chicago: Univ. of Chicago Press, 1959).

border, Puno, is 80 percent Indian and many of these, as noted above, migrate to the city of Arequipa. Thus there is a noticeable aboriginal influence in all phases of Arequipenian life, including agriculture.

The trinity of basic values of the highland Indian are the family, land, and animals.[1] As Fried states, "the family's survival has over-riding priority over any wider community or, at times, even kinship considerations . . . [and so] it is extremely difficult to get Indians to work when the motive is not personal welfare."[2] Continuing, he states that "the Vicosino dodges and maneuvers to avoid communal works, such as clearing irrigation channels and road repairs."[3] In sustaining the family, land is valued as an important element in the hierarchy of social status.[4] Nevertheless, the Indian feels subordinate to "the periodic changes that mark the stations of nature and animals, which is expressed by the veneration of nature based on the psychic interpretation of natural phenomena."[5] Escobar further relates that this view of nature is expressed in an hierarchical manner, so that hills are seen as grandparents (apus or awkis) and the land as mother earth (Pacha Mama). Other elements in the landscape such as rivers and creeks are believed to possess spirits that also demand reverence. Thus the Indian, having great veneration for the earth, perceives a close symbolic link with his land.

The Indian also treats the products of the land with a reverence which is manifested in ceremonies that occur throughout the year.[6] An example of offerings made by the Indians to the deities and spirits was witnessed and participated in by the author near Juliaca in the Department of Puno. An old tradition being practiced by a group of Indians had to do with the raising of potatoes. At the beginning of the ceremony all of the males went to the nearby potato field to bury apples among the young plants. When the apples were all buried, the men, together with the women, seated themselves in a large circle around the old matriarch of

---

[1]Jacob Fried, "Social Organization and Personal Security in a Peruvian Hacienda Indian Community: Vicos," American Anthropologist 64 (August 1962): 774.

[2]Ibid., p. 779.

[3]Ibid.

[4]Gabriel M. Escobar, Organización Social y Cultural del Sur del Perú, Serie: Antropología Social, no. 7 (México, D.F.: Instituto Indigenista Interamericano, 1967), p. 106.

[5]Ibid., p. 83.

[6]Ibid., p. 106.

the group, who sprinkled pisco or aguardiente over a small pile of coca leaves. After wetting the leaves, the elderly lady began to select the finer leaves and meditate over them. This ritual was performed with the hope that the potatoes would grow as large as the apples planted. The Indians' belief in the efficacy of magic was evident. The group was also displaying a pragmatic view of the world, in that these apples would in time serve as a form of ferilizer for the crop.

According to Escobar, the Indian's vision of the world has been static in recent generations, with very little susceptibility to change.[1] Before the colonial period, in the time of the Incas, the Indian seemed to have a more active attitude toward the land, in that modifications to the landscape in the form of terracing and irrigating new land were more frequently carried out.[2]

One trait that persists up to the present is the work ethic. Indians hold high regard for work; idleness is looked upon as a sin.

Two conditions have brought a feeling of insecurity to many Indian farmers. One of these has been the loss of their land during the colonial period, through the institution of the encomienda whereby groups of Indians and their land were put under the control of selected Spaniards. After the demise of this institution, the land rarely reverted to Indian ownership. Another condition that intensifies the insecurity among the farming group is the small size of holdings, which may consist of as little as one or two furrows. This is especially true in the Lake Titicaca Basin. In a situation of this kind, the Indian is not likely to gamble on new farming methods. Because of these conditions, the Indian has developed a negative attitude of seclusion or retreat and a distrust of other social classes.[3] There is a lot of discontent and little desire to make contact with the world outside the Indian community.[4]

Cholos

Whereas the lower class in the sierra of southern Peru, as in the Department of Puno, consists of Indians, the bulk of the lower class in the Department of Arequipa consists of Cholos - Indians who have become members of the national community. In Peru, as in other countries of Latin America, the Indian may readily move into the national community provided he transforms his life in three

---

[1]Ibid., p. 82. [2]Ibid., p. 85. [3]Ibid., p. 106.

[4]Schaedel, La Demografía y Los Recursos Humanos del Sur del Perú, pp. 25-26.

areas. He must learn to speak the national language, adopt European-type clothing, and move away from the Indian community.[1] Many of the laborers in the Chili Basin, especially farm laborers, are former Indians who now might be classed as Cholos. This group maintains the values of Indian heritage as well as those of the middle and even upper class groups.[2] This dichotomy of heritage is reflected in a number of ways. Among the farming group of the Arequipa area, the Arequipeño at a social gathering may pay his due respects first to Pacha Mama before he begins to drink, by pouring a small quantity of his beverage on the ground. It is also not uncommon, as a Christian reflection, to see crosses in a field of onions, placed there with the hope of assuring a good harvest. In referring to the Cholos, Schaedel observed that, "among the total population of the south, [this group is] the most disposed to accepting and initiating change provided it can yield some advantage."[3]

Middle Class

Even though the middle class constitutes a small percentage of the total population in the Chili Basin, it is influential in that its members occupy so many of the leadership posts, especially in agriculture. They may be owners, administrators, or renters of farms, or government employees, especially engineers, within agricultural agencies. Also according to Gillin, the middle class has an importance far beyond that indicated by its size because, as is the case in southern Peru, it is the group "most in touch with the modern world; most susceptible to influences for change; and most potent in internal affairs of their own nation."[4]

The middle class differs from the lower class in that it disdains manual labor. The middle class, in fact, "regards manual work as a pastime or distraction of short duration, as leisure, not work that is habitually engaged in."[5] This attitude toward manual work has been inherited from colonial times; "the value of manual work is negative and for that reason any person engaged

---

[1] John P. Gillin, "Some Signposts for Policy," in Social Change in Latin America Today, eds. Richard N. Adams et al. (New York: Vintage Books, 1960), p. 21.

[2] Schaedel, La Demografía y Los Recursos Humanos del Sur del Perú, p. 40.

[3] Ibid., p. 41.

[4] Gillin, "Some Signposts for Policy," p. 26.

[5] Escobar, Organización Social y Cultural del Sur del Perú, p. 105.

in this type of endeavor cannot be considered decent people (gente decente)."[1] Also there is often little incentive for performing efficient work; for example, in a state agricultural agency where both nepotism and the institution of godparenthood (compadrazgo) are operating.[2]

Another marked difference between the middle class and the lower class, especially the Indian, is in the investing of the earth with spiritual powers. The middle class looks upon land less reverently; land can be manipulated and controlled to any degree for agricultural purposes by man himself.[3]

Another trait of the middle class that has relevance in the development of agriculture is that scientific, creative endeavors have received little consideration. In fact, as Gillin states, "the manipulation of things to explore mechanical or functional potentialities is not of great interest to them."[4] Gillin goes on to say that "most Latin Americans of the middle class are not dominated by a 'mechanistic' world view, nor are they usually moved by an urge to undertake systematic, empirical investigations."[5] Gillin also observes that "this class is not as enamored of utilitarianism or pragmatism as we in the United States."[6]

Another important value to consider in understanding the middle class is their sense of fatalism. This might be expressed as passive resignation in that if one has done his best and fails, it is due to the ways of fate and not his personal fault.[7]

Upper Class

The upper class in South Peru, in particular the Chili Basin, consists of a minute group of individuals that may include, among others, project engineers in resource development programs. Hacendados who reside in the city of Arequipa but own large agricultural holdings (haciendas) in the Department of Puno may also be in this group. This class shares many of the values held by the middle class. A characteristic, however, that distinguishes the upper class from the middle class is that the former's power is based on ancestry and possibly great wealth.[8]

---

[1] Schaedel, La Demografía y Los Recursos Humanos del Sur del Perú, p. 46.

[2] Ibid., p. 44.

[3] Escobar, Organización Social y Cultural del Sur del Perú, p. 85.

[4] Gillin, "Some Signposts for Policy," p. 40.

[5] Ibid. [6] Ibid., p. 42. [7] Ibid., p. 47. [8] Ibid., p. 25.

In some instances, especially among engineers and other public employees who are not of the local area, there is a tendency, when the opportunity presents itself, to gravitate back to the capital of the country, Lima. Some of the more capable personnel in governmental service in the Chili Basin see their residence in the area as but temporary. Thus strong technical leadership may be absent from the various agricultural agencies, and continuity of programs may be difficult to achieve in this region. This attitude of impermanence is countered by an opposite attitude generally possessed by many of the engineers of local origin. Regionalism is very strong in South Peru and especially in Arequipa.[1] According to Escobar, South Peru has the highest regional spirit of any area in the country.[2] This great patriotic fervor for the local area is well expressed in the final act of the regional poet, Mariano Melgar. In defending his homeland from outsiders, he died doing so.[3]

The regional viewpoint of the politicians who began their careers in the Department of Arequipa has influenced policies throughout the entire country. In fact, many revolutions in Peru have started here in Arequipa. At one point in his presidential career, Fernando Belaunde Terry took to the streets of Arequipa to defend his right to the presidency. This act of the president reflects the suspicions of the populace of any covert operations that emanate from Lima.

This strong regional sentiment can result in negative reactions. Regionalism could foster reluctance on the part of the *Arequipeño* to accept technology and ideas introduced from the outside.

Human Geography and the Use of Irrigation Water

Irrigated agriculture has been practiced in the Chili Basin for over four centuries. The familiarity and long experience of the people with this area allowed a viable agricultural system to evolve. A key element in their program has been the use of both land and water resources. A concern for these scarce resources, especially in the use of irrigation water, is reflected in the cultural, social, and economic history of the area.

Several elements, such as the settlement patterns, have affected the use of irrigation water. At first glance, it would

---

[1] Escobar, *Organización Social y Cultural del Sur del Perú*, p. 88.

[2] Ibid.

[3] José Carlos Mariátequi, *7 Ensayos de Interpretación de la Realidad Peruana*, 3rd ed. (Lima, Perú: Empresa Editora Amauta S.A., 1952), pp. 281-284.

appear that farmers who live on their farms would give closer attention to the management of water than those living in farm villages away from their farms. But where social and other amenities are not obtained in on-farm living, farmers have tended to move to the cities. Many farm houses located in the country are occupied, not by the owner's family, but by an overseer who has less concern than the owner for the effectiveness of all farm operations, including the management of water.

The rapid growth of population in the Chili Basin in the vicinity of the city of Arequipa has two obvious effects on the use of irrigation water. As the city expands outward onto once-irrigated lands, the change in land use results in a lower consumptive use of water. Yet, with the growing demands for both jobs and food, many of the leaders in the area see the concomitant need for more irrigation projects. These new projects will be more costly per hectare than previous ones. The more accessible and richer lands are put under the ditch first. Besides being more costly to build, the newer projects can be more difficult to operate and maintain. If not properly managed, the sandy soils present in some of the contemplated projects can result in an inordinate waste of water.

Intensive agriculture is synonymous with irrigated agriculture in the Department of Arequipa. In fact, all land in commercial agriculture in the Chili Basin is irrigated. With alfalfa the dominant crop in the study area, farmers have wide latitude in disciplining themselves in the wise use of irrigation water.[1] Alfalfa can sustain a wide range of water practices that many other crops in the region cannot.

The nature of farming and farmers in this area almost ensures an agricultural setting that does not change rapidly. In part, availability of opportunities is the cause of this dilemma. Some innovations in agricultural practices have been accepted, but it is likely that improvements in the use of irrigation water will be slow and difficult. With many small fragmented farms in the Chili Basin, farmers cannot always give proper attention to water use. Not only is the local community extremely cautious about accepting new ideas and techniques from the scientific community, but farmers do not have access to, or seek out, information, especially on water use.

The perceptions of the environment held by various groups

---

[1]An important related observation made by Maass and Anderson is, "To succeed for any length of time to capture and distribute available water, and to control the amount of land placed under irrigation, farmers must develop self-discipline and a high level of community organization." Maass and Anderson, . . . and the Desert Shall Rejoice, p. 1. (Emphasis mine.)

in the area also influence the management of irrigation water. The local community acts as a conservative counterbalance to the desires of change agents from outside, with radically different ideas about how best to manage resources. As a result, conflicts may arise in the use of water. Not only the highly trained scientist or technician but also the highland Indian from the Lake Titicaca Basin with little formal schooling is able to influence local agricultural practices. As Wagner observes, "In the Andean highlands, potatoes, quinoa, ulluco, oca, and other peculiar vegetables are grown (by a form of cultivation known as 'horticulture')."[1] A pertinent point made by Wagner about horticulture is that the individual plant is given much hand care. As will be shown later, irrigators from the Lake Titicaca Basin seem at times to practice a variant of horticulture in using irrigation water.

---

[1] Philip L. Wagner, The Human Use of the Earth (Glencoe, Illinois: The Free Press of Glencoe, 1960), p. 111.

CHAPTER III

THE PHYSICAL ENVIRONMENT AS MODIFIED BY IRRIGATED AGRICULTURE

The Chili Basin offers both resource opportunities and problems which affect the use of irrigation water. A range of physical factors including topography, soils, climate, and water supply set the stage for irrigated agriculture. That the physical environment is important is understandable, but the man-made modifications of this environment must not be overlooked in appreciating the interrelationships that exist between the physical milieu and its use. Therefore, special attention is given to the modifications made in the water supply and the procedures followed in its allocation. The following sections describe and explain the physical environment as modified by irrigated agriculture. This is useful as a basis for understanding the management of irrigation water as presented in the next chapter.

General Description

Though much of the coastal irrigated land is in northern and central Peru, the south coast has several oases of local and national importance, a number of them in the Quilca River Basin. Two of these, found in the Campiña of Arequipa (Campiña) and on the Pampa of La Joya within the Chili Basin, lie approximately midway along the Quilca River (figures 1 and 2).

From its headwaters in the Andes, which average 4,000 m in elevation, the Quilca River passes through a diversity of landscapes. On flowing out of the Andes, it crosses the Campiña, located about 2,500 m above sea level; cuts through the Caldera Range which has peaks over 1,400 m; and finally empties into the Pacific Ocean at the small village of Pueblo Nuevo.

This river system connects the Campiña with the Pampa of La Joya. The Campiña is triangular in shape (figure 2). The northwest side roughly parallels the Yura River, the south and northeast sides converge near the village of Sogay. The Andes, with volcanic peaks over 5,000 m, face the Campiña on the northeast side. The Caldera Range delimits it on the south and a portion of the northwest. The

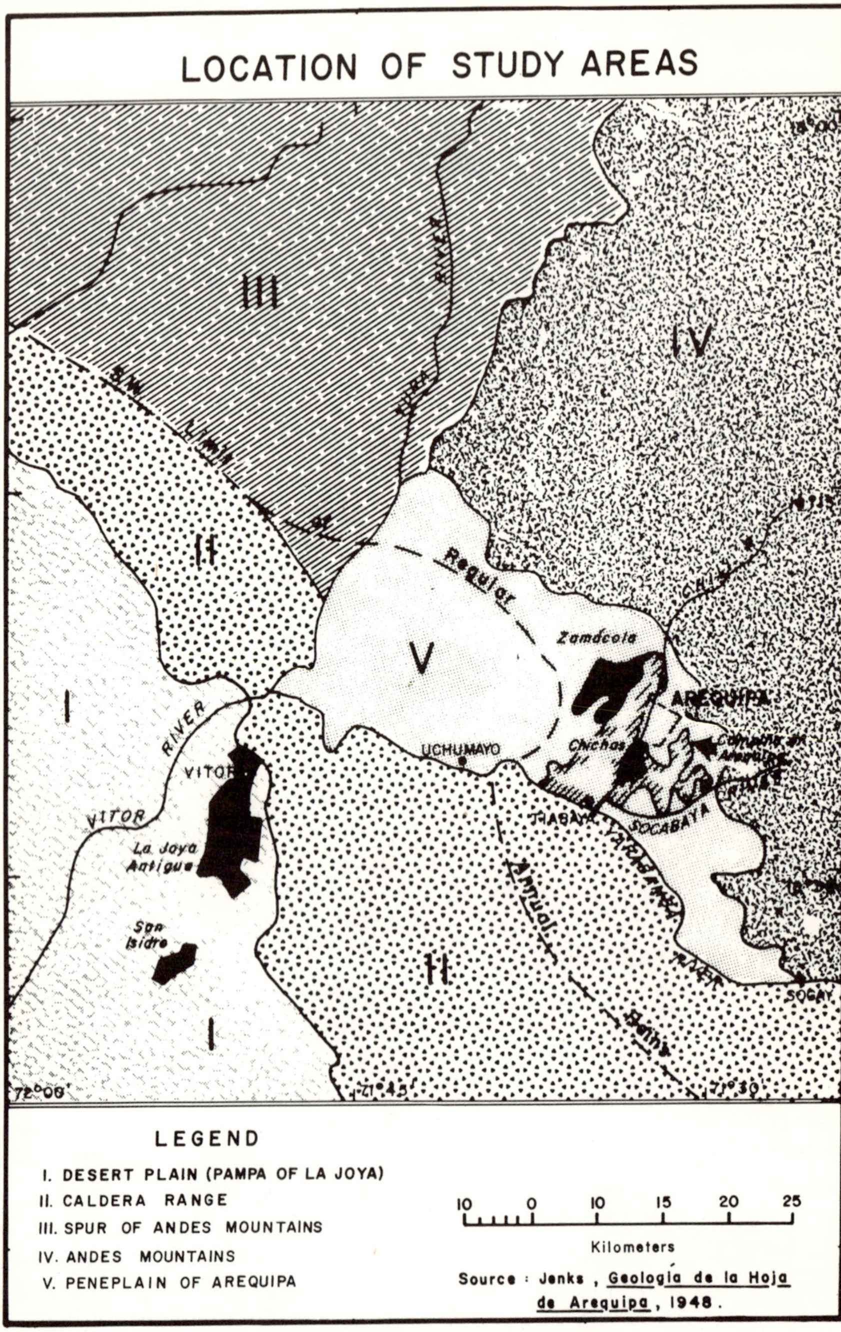

Fig. 2 - Location of Study Areas.

remaining portion on the northwest is demarcated by a spur of the Andes, completing the enclosure of the Campiña.

While much of the Campiña is underlain by volcanic tuff, it is only in the extreme northwest portion that this is exposed at the surface; the remaining area is overlain by alluvium and mud flows.[1] It is in this latter sector that the initial thrust of irrigated agriculture took place. Some of the older irrigated areas are found on the natural terraces bordering the Chili River, and many man-made terraces allow for maximum use of hillsides and steeply sloping land. Scattered over the Campiña are remnants of intrusive plutonic rocks, which stand out as small rugged hills. In adjusting to the limited amount of good cultivable land, many of the small villages were built on these igneous outcrops, as noted in the previous chapter.

Unlike the Campiña, irrigated agriculture came late to the Pampa of La Joya. In the Campiña the Chili River has not cut deeply into the bedrock making it relatively easy to divert river water for irrigation purposes. One reason for the delay in bringing irrigation to the coastal pampa is the difficulty in diverting water of the Chili River, which at that point flows through a canyon 250 m below the surface of the coastal plain. This pampa has a relatively flat surface that slopes gently toward the southwest. It averages 1,600 m above sea level near the Caldera Range and 1,100 m on abutting the Coastal Range. The pampa is underlain by semiconsolidated tertiary sediments of soft sandstone, shale and gravel known as the Monquegua Formation, in beds more or less parallel to the surface.[2] On this desert plain a braided intermittent stream system emerges from the Caldera Range. A noticeable feature of the landscape of this pampa is the large number of barchans moving in bands toward the northeast.[3] These dunes clog irrigation ditches and obstruct road traffic; they also affect farm operations, necessitating unaccustomed adjustments in the use of irrigation water.

---

[1] Clarence N. Fenner, Geología de los Alrededores de Arequipa, trans. Carlos Nicholson (Arequipa, Perú: Editorial La Colmena, S.A., 1940); William F. Jenks, "An Account of the Geology of the Arequipa Region," The Peruvian Times, February, 1946, pp. 25-31; William F. Jenks, Geología de la Hoja de Arequipa al 200,000 (Boletín no. 9; Lima, Perú: Instituto Geologico del Perú, 1948); Luis V. Vargas, Geología del Cuadrángulo de Arequipa, (Boletín no. 24; Lima, Perú: Servicio de Geología y Minería, 1970).

[2] Jenks, "An Account of the Geology of the Arequipa Region," p. 25; Jenks, Geología de la Hoja de Arequipa al 200,000, pp. 97-98; Vargas, V., Geología del Cuadrángulo de Arequipa, p. 12.

[3] Jenks, "An Account of the Geology of the Arequipa Region, p. 25; H.J. Finkel, "The Barchans of Southern Peru," Journal of Geology 67 (1959), 614-647; Katharina Lettau and Heinz Lettau, "Bulk Transport of Sand by the Barchans of the Pampa de La Joya in Southern Peru," Zeitchrift für Geomophologie, n.f. 13 (1969), 182-195.

Soils

Generalized soil maps (figures 3 and 4) show the texture and depth of soils in the Campiña and on the Pampa of La Joya within the irrigated areas.[1] In a general way these maps reflect two important physical characteristics of soil that affect the operation of irrigation schemes and in particular the use of water: permeability and moisture retention. While there are many factors influencing the permeability of a soil, Houk states that:

> . . . (soil) permeability) is largely dependent on soil texture and structure. . . .
>
> When the pore spaces in a soil are relatively large, as in sandy soils of coarse texture, the passage of moisture is facilitated and soil has a high degree of permeability. When the pore spaces are relatively small, as in clay soils of fine texture, the passage of moisture is retarded and the soil has a low degree of permeability.[2]

Moisture retention (total available moisture) refers to the water held by a soil between field capacity and a preselected depletion limit.[3] Moisture retention is affected by soil texture. Even though there is a wide range in the moisture characteristics of soils within each soil type,[4] it can be shown that there is a relation between total available soil moisture and texture: the lighter or sandier the soil, the lower the amount of total available moisture.[5]

An additional factor important to irrigation water use is the capacity of the soil to store water, which can be greatly affected if there is a shallow wetting zone.[6] This wetting zone (depth of

---

[1]Basic studies of the soils in the Chili Basin are Equipo Agropecuario Peruano-Israel, "Estudio de Suelos, Preliminar, Colonización La Joya," Arequipa, Perú, 1967, (Mimeographed); Equipo Agropecuario Peruano-Israel, Proyecto de Planificación, Colonización "La Joya" (Arequipa, Perú: Programa Conjunto OEA-Israel, 1967), pp. 48-54; República del Perú, Oficina Nacional de Evaluación de Recursos Naturales (ONERN), Inventario, Evaluación y Uso Racional de los Recursos Naturales de la Costa: Cuencas de los Rios Quilca y Tambo, 3 vols. (Lima, Perú: ONERN, 1975), vol. 1, pp. 103-346 and vol. 3, pp. 122-81.

[2]Ivan E. Houk, Irrigation Engineering, 2 vols. (New York: John Wiley and Sons, Inc., 1951), vol. 1, p. 25.

[3]M.G. Bos and J. Nugteren, On Irrigation Efficiencies (Wageningen, The Netherlands: International Institute for Land Reclamation and Improvement, 1974), pp. 2-3.

[4]Ray K. Linsley and Joseph B. Franzini, Water Resources Engineering (New York: McGraw-Hill Book Company, 1964), p. 387.

[5]Orson W. Israelson and Vaughn E. Hansen, Irrigation Principles and Practices, 3rd ed. (New York: John Wiley and Sons, Inc., 1962), p. 168.

[6]G. Varallyay, "Soil Water Problems Related to Salinity and Alkalinity in Irrigated Lands," in Worthington, Arid Land Irrigation in Developing Countries, p. 262.

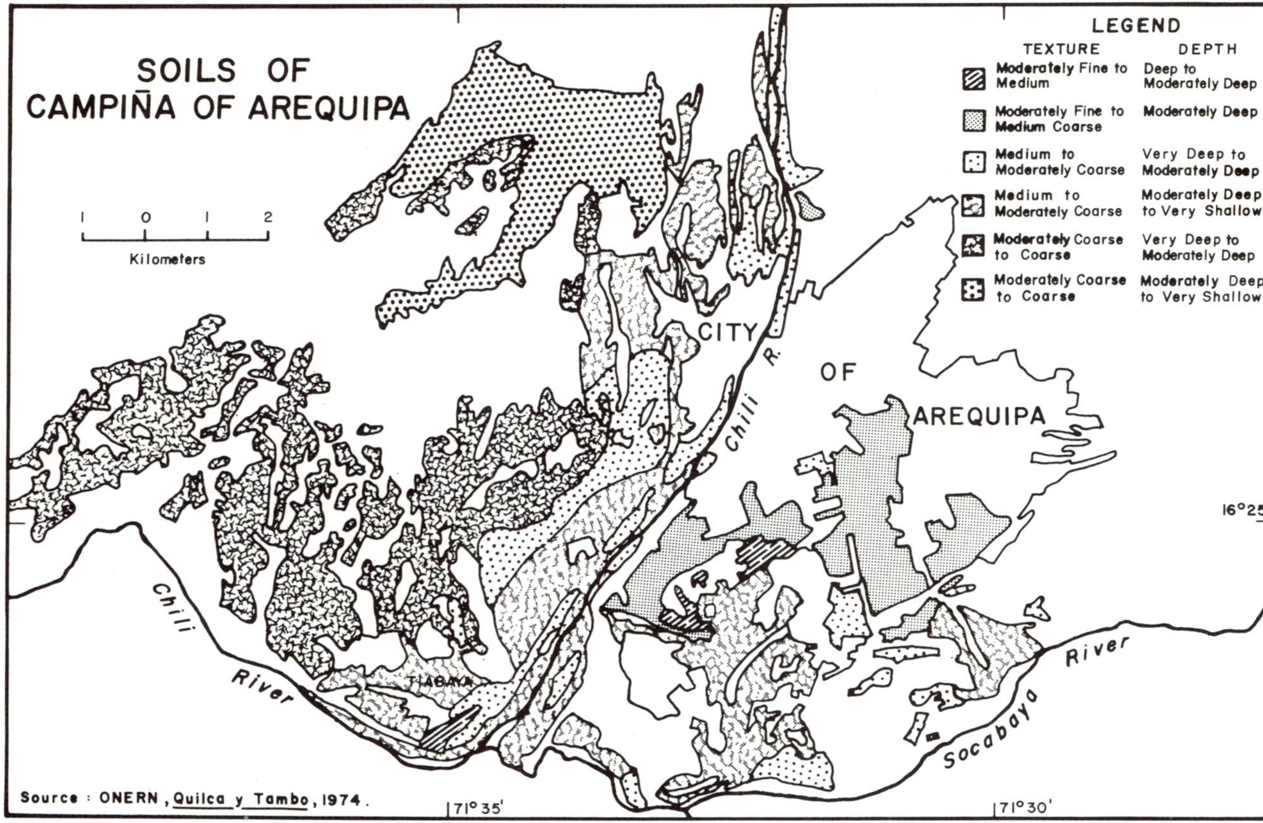

Fig. 3 - Soils of Campiña of Arequipa.

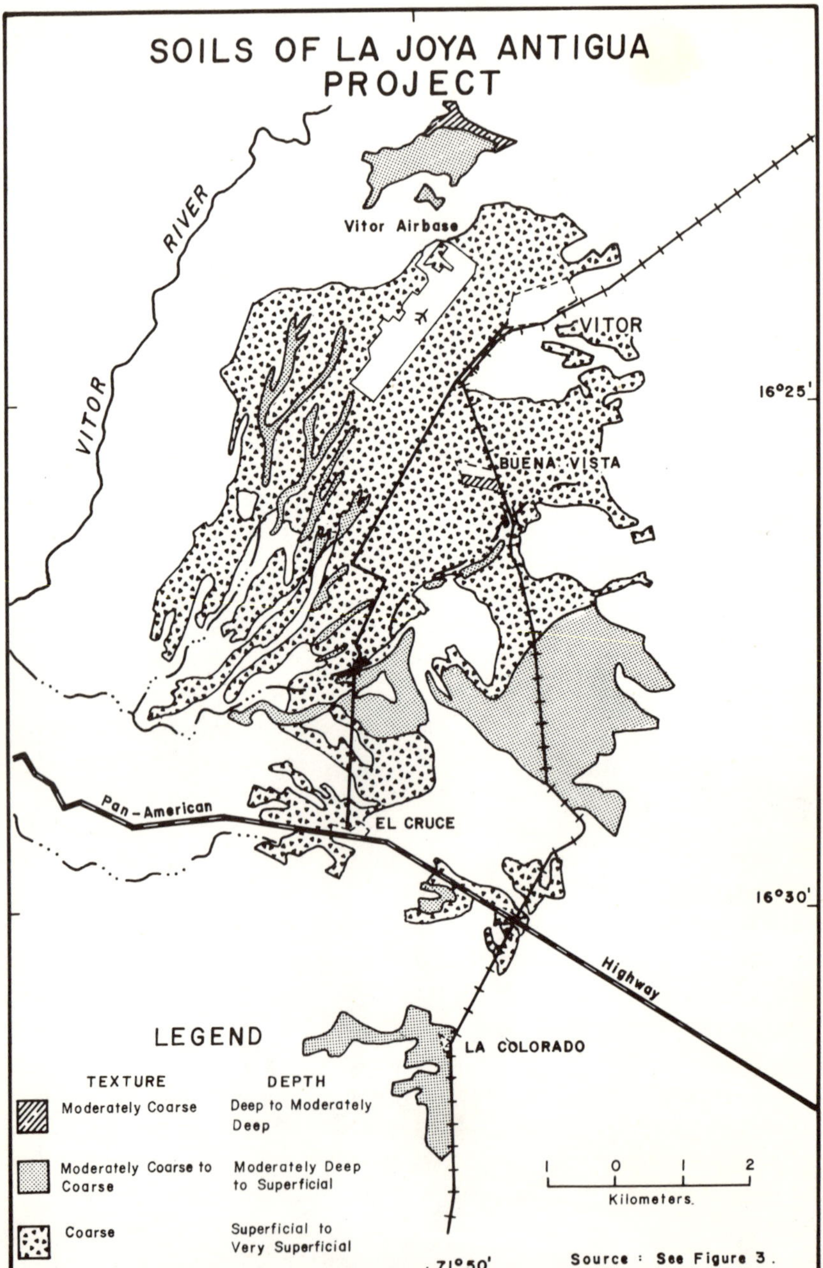

Fig. 4 - Soils of La Joya Antigua Project.

soil) is shown in figures 3 and 4.

While most of the soils in the Chili Basin are alluvial, they are finer and deeper in the Campiña than on the Pampa of La Joya. Yet there are variations within each area. In the Campiña the soils southeast of the Chili River tend to be more finely textured and deeper than those found to the northwest of the river. Whereas some of the coarser and shallower soils in the Campiña are in the Zamácola Irrigation Project, those soils located near the Chili River on the nearby terraces are quite fine and rather deep. This is especially the case near Tiabaya. Northwest of Pampas Nuevas, where the soils are coarse, erosion is epitomized by the deepening of the beds of irrigation canals.

The soils of the La Joya irrigation projects are moderately coarse to coarse, with a shallow profile. Surprisingly, in the lower-lying areas of the original project, small tracts of land have become waterlogged. Nevertheless, despite the limited waterlogged zones, the soils are sufficiently coarse to be well-drained.

For efficient application of irrigation water in sandy, shallow soils, it is necessary to apply water frequently and in small amounts. Neither practice may be feasible due to a combination of factors such as the rigidity in the water delivery system employed, the physical constraints stemming from the irrigation methods used, or the overriding influence of the perception of the situation held by farmers. In the coarser soils, farmers may not be able to readily appreciate the damage occurring in over-irrigating, if waterlogging does not occur immediately. Another management problem in using these sandy, shallow soils is their great susceptibility to erosion. While farmers may be aware of what is happening, they may be helpless or at least reluctant to prevent this damage, due to a unique set of circumstances. Using too large a stream flow at any one time, where the fields have pronounced longitudinal slopes, hastens soil erosion.[1] Even though field erosion does occur in areas containing finely textured soil, the loss of soil is not as evident, since the deeper profiles do not readily show gross signs of erosion.

In summary, the Campiña possesses an appreciable amount of ideal soils, both in terms of good permeability, and in ability to store a relatively large amount of readily available moisture for plant use. In part this is reflected by the intense use made of this land for such specialty crops as onions, garlic, and other vegetables, as noted in the previous chapter. In contrast, some

---

[1] Stream flows used in irrigating fields and corresponding slopes of fields are assessed in chapter IV.

agricultural specialists have likened agriculture on the Pampa of La Joya to a modified hydroponic system.[1] Frequent applications of both water and fertilizers are supplied to sandy soil that has as one of its chief functions the physical support of the plant in an otherwise barren environment.

The Water Supply

The various sources of water in the study area are precipitation, surface flow, and ground water, only the second one being significantly important. Precipitation is meager and streamflow is erratic, both annually and seasonally. Groundwater is largely untapped due to governmental sanction, which prevents possible changes in surface flows. The modifications affecting surface supplies are storage facilities and interbasin transfers.

Climate

"Nowhere else on the earth is there a coastal desert of such intense aridity which spans so many degrees of latitude as the Chilean-Peruvian desert."[2] The six principle factors making for aridity on this coast are:

1. a strong and positionally stable anticyclonic cell
2. abnormally cool coastal waters
3. an abrupt mountainous coast
4. a coastline which continues to bend westward into the circulation of the open ocean until near the equator
5. an abrupt termination of the anticyclonic circulation at the coastline
6. an unusually strong sea breeze.[3]

Resulting from this stable anticyclonic cell is a presistent thermal inversion, which is intensified by the contact the surface air makes with the cool upwelling Peruvian Current. In addition to abruptly checking the eastern end of the anticyclone on the west coast of South America, the Andes act as an impediment in preventing the interchange of the tropical humid air east of the Andes to move

---

[1] La Dirección de Aguas de Regadío del Ministerio de Agricultura, Arequipa interviews with employees, 1968.

[2] Glenn T. Trewartha, The Earth's Problem Climates (Madison: Univ. of Wisconsin Press), 1961, p. 22.

[3] Ibid., p. 31.

along the west coast of Peru.[1]

A recent theory by Lettau on the cause of the coastal arid condition expands on earlier observations. He noted the importance of the vertical structure and heat budget of the air masses over the cold coastal waters and the adjacent land masses.[2]

Straddling this arid zone is the Chili Basin (figure 1). The Pampa of La Joya lies entirely within this extremely arid desert, while the Campiña lies on its border (figure 2). As Johnson observed, "the southern city of Arequipa . . . appears as an oasis in an otherwise barren landscape, due to irrigation, not to local rainfall."[3]

With the increase in elevation, especially above the city of Arequipa, aridity diminishes. Thus in the upper basin of the Quilca at an elevation above 2,800 m, a more humid environment exists.[4] Johnson correctly observed that, "Above the City (of Arequipa) there does not appear to be any great increase in precipitation; the dull brown landscape continues up to the snows of the 6,000 m Chachani."[5] Yet the humid environment is more noticeable in the high sierra of the upper basin of the Quilca, to the north and east of the mountain peaks and El Misti.

Temperature and Sunshine

While the temperatures within the Chili Basin are moderate, extreme conditions do occur, especially in the upper basin of the Quilca. The average monthly temperature on the Pampa of La Joya ranges from 17°C to 18°C, while at a higher elevation in the Campina,

---

[1] United Nations, Economic Commission for Latin America, "Los Recursos Hidráulicos de America Latina: Peru," E/CN. 12/794, July 31, 1968, p. 24 (mimeo).

[2] Heinz Lettau, Dynamic and Energetic Factors which Cause and Limit Aridity along South America's Pacific Coast," in Climates of Central and South America, ed. Werner Schwerdtfeger (Amsterdam: Elsevier Scientific Publishing Co. 1976), pp. 188-192. For additional comments on the causes of aridity along the Peruvian coast see Paul E. Lydolph, "On the Causes of Aridity along a Selected Group of Coasts," pp. 67-72 and Federico J. Prohaska, "New Evidence on the Climatic Controls along the Peruvian Coast," pp. 91-107 in Coastal Deserts, Their Natural and Human Environments, ed. David H. Amiran and Andrew W. Wilson (Tuscon: The University of Arizona Press, 1973).

[3] A. M. Johnson, "The Climate of Peru, Boliva and Ecuador," in Schwerdtfeger, Climates of Central and South America, p. 153.

[4] Oficina Nacional de Evaluación de Recursos Naturales (ONERN), Inventario, Evaluación y Uso Racional de los Recursos Naturales de la Costa: Cuencas de los Rios Quilca y Tambo, III vols. (Lima, Perú: ONERN, 1974), vol. I, p. 409.

[5] Johnson, "The Climate of Peru, Bolivia and Ecuador," p. 165.

the range is from 14°C to 17°C (table 15). On the other hand, in the upper basin of the Quilca, at Imata, the average monthly temperature ranges from -0.3°C to 5°C, much lower than in the Chili Basin (table 15).

Periods of extreme temperatures occur in the Chili Basin, but it is possible to grow crops year-round. The highest temperatures are on the Pampa of La Joya, with a monthly mean maximum as high as 32.7°C in August. The coldest temperatures are in the Campiña, with a monthly mean minimum as low as 0.5°C. In this latter area occasional frost occurs in the winter months, when the temperature can drop one or two degrees below 0°C.[1] Even colder temperatures exist in the upper basin of the Quilca. Here the monthly mean minimum temperature is always below 0°C and on many occasions it is much lower. The lowest monthly mean minimum temperature on record, -16°C, occurred in June, 1950. Low temperatures naturally preclude year-round cropping in this area. However, hardy natural grasses, such as ichu (Stipa ichu), provide forage for cattle, llama, alpaca, and sheep, and material for thatching.

In the study area sunlight is not a limiting factor for agriculture. In studying the climate of Bolivia, Ecuador, and Peru, Johnson stated that:

> the most sunny area is probably the western Andean slope of southern Peru, above the coastal inversion but below the rainy regions of the high sierra. (The city of) Arequipa is in this optimum area where very high solar radiation values are recorded . . . (The) annual sum of global radiation, 228 Kcal/cm (for the city of Arequipa) is equal to that of much of the central Sahara of north Africa.[2]

At the La Pampilla station, in the Campiña, the average daily duration of sunlight is approximately 9 hours, compared to 10 hours on the Pampa of La Joya (table 15). Sunlight is slightly less during the summer season for both areas. The decrease in sunlight on the Pampa of La Joya is conditioned by occasional overcast skies, while in the Campiña, the summer rains are responsible for a lesser amount of sunlight.[3] The upper basin of the Quilca has less sunlight than

---

[1] Joseph A. Tosi, Jr., Zonas de Vida Natural en el Perú (Boletín Técnico, no. 5. Lima, Peru: Instituto Interamerican de Ciencias Agrícolas de la OEA, Zona Andina, 1960), p. 38.

[2] Johnson, "The Climate of Peru, Bolivia and Ecuador," pp. 183, 185.

[3] Johnson though shows that more solar radiation reaches the city of Arequipa in summer than winter. See Johnson, "The Climate of Peru, Bolivia, and Ecuador," table XII, p. 183.

the Chili Basin. The average daily duration of sunlight is only 7 hours, with the minimum daily average occurring during February, the height of the rainy season (table 15).

Precipitation

As a source of water for crop production, precipitation is pertinent. The Chili Basin receives only a small amount of precipitation annually, and its occurrence is erratic. The average annual precipitation for the Pampa of La Joya is 23 mm, while in the Campiña it is 101 mm (table 15). These two areas stand out in marked contrast with the upper basin of the Quilca, where the average annual precipitation is 557 mm (table 15). The upper basin also has a smaller yearly variation in precipitation; the coefficient of variation is only 32 at Imata, compared with approximately 104 for the Chili Basin.

Another measure of variation in precipitation is the mean deviation indicated for the two stations in the Chili Basin and the one station in the upper basin of the Quilca (table 16). Within the study area the mean deviation as a percentage of the mean is 74 percent, much higher than the 23 percent at Imata. Table 16 also shows the quartile positions for these stations. The range in precipitation in the Chili Basin is between 380 percent and 460 percent of the mean. With a marked decrease in variability in the upper basin, the range is 143 percent of the mean.

Throughout the Chili and Upper Quilca Basins there is a definite seasonal regime in precipitation. The Chili Basin has a pronounced "rainy season," extending from January through March. Of the total average annual precipitation, the Pampa of La Joya and the Campiña receive 84 and 94 percent, respectively, during the first three months of the year (tables 17 and 18). The remaining portion of the year is generally very dry for both areas. In about half of the years of record, there was no precipitation during the extensively dry period from April through September (tables 17 and 18).

The rainy season of the upper basin of the Quilca is longer than in the Chili Basin (table 15). Even though the upper basin has a dry season, precipitation normally occurs during all months; however, almost 65 percent of the yearly average falls during the first three months of the year, and another 23 percent from October through December (table 19).

The length of periods of continuous drought reflects in part the aridity of an area. The period of drought varies throughout the Chili and Quilca Basins (table 20). The persistance of drought

TABLE 15

CLIMATIC DATA FOR FOUR STATIONS IN THE QUILCA BASIN

Station	Data	Unit	Period of Record	Oct.	Nov.	Dec.	Jan.	Feb.	Mar.
Vitor La Joya Pampa elev. 1, 542 m lat. 16°28'S	temperature	°C	1935-1968	18.1	18.0	18.2	18.0	18.2	18.4
	precipitation	mm	1935-1968	0.3	0.8	0.8	7.2	8.8	3.5
	sunshine	hrs/day	1939-1958	10.2	10.3	10.0	9.2	9.7	9.9
Campiña of Arequipa (CORPAC) elev. 2,525 m lat. 16°21'S	temperature	°C	1949-1971	16.8	16.9	16.7	16.0	15.6	15.8
	precipitation	mm	1949-1971	0.2	0.8	4.4	30.1	46.0	17.7
Imata Upper Quilca elev. 4,450 m lat. 16°10'S	temperature	°C	1936-1971	3.2	3.9	4.8	5.3	5.5	5.3
	precipitation	mm	1936-1971	15.2	26.2	80.7	134.2	125.6	95.9
	sunshine	hrs/day	1952-1967	8.1	7.7	5.8	5.5	5.0	5.8
La Pampilla Campiña of Arequipa elev. 2,350 m lat. 16°26'S	sunshine	hrs/day	1962-1970	9.8	10.0	8.8	7.3	6.4	7.5

TABLE 15--Continued

Station	Data	Unit	Period of Record	Apr.	May	June	July	Aug.	Sept.	Year
Vitor										
La Joya Pampa	temperature	°C	1935-1968	17.9	17.9	17.4	17.2	17.5	18.0	18.0
elev. 1,542 m	precipitation	mm	1935-1968	0.6	0.3	0.1	0.1	0.0	0.4	23.2
lat. 16°28'S	sunshine	hrs/day	1939-1958	10.1	10.0	10.3	10.3	10.4	10.3	10.1
Campiña of Arequipa										
(CORPAC)	temperature	°C	1949-1971	15.5	15.2	14.3	14.3	15.0	16.3	15.7
elev. 2,525 m	precipitation	mm	1949-1971	0.5	0.4	0.0	0.0	0.0	0.8	100.9
lat. 16°21'S										
Imata										
Upper Quilca	temperature	°C	1936-1971	4.2	2.1	0.5	-0.3	0.5	2.3	3.2
elev. 4,450 m	precipitation	mm	1936-1971	33.5	12.0	3.9	1.9	3.3	13.0	556.6
lat. 16°10'S	sunshine	hrs/day	1952-1967	7.3	8.5	9.4	9.3	9.0	7.9	7.4
La Pampilla										
Campiña of Arequipa	sunshine	hrs/day	1962-1970	8.6	9.4	9.6	9.5	10.0	9.7	8.9
elev. 2,350 m										
lat. 16°26'S										

SOURCE: Computed from records of ONERN.

TABLE 16

RAINFALL: MEAN AND QUARTILE POSITION AT THREE STATIONS

	Vitor 30 years		Arequipa (CORPAC) 22 years	
	Precipitation (mm)	Percentage of mean	Precipitation (mm)	Percentage of mean
Lower Extreme	1.0	3.9	9.6	9.5
Lower Fourth	8.4	32.6	30.9	30.6
Mean	25.8	100.0	100.9	100.0
Upper Fourth	33.6	130.2	150.0	148.7
Upper Extreme	100.1	388.0	469.7	465.5
Mean Deviation as Percentage of Mean Rainfall	74.0		74.3	
Range	381		460	

TABLE 16--Continued

	Imata 31 years	
	Precipitation (mm)	Percentage of mean
Lower Extreme	278.1	51.7
Lower Fourth	426.3	79.2
Mean	538.1	100.0
Upper Fourth	584.4	108.6
Upper Extreme	1,049.1	195.0
Mean Deviation as Percentage of Mean Rainfall	22.9	
Range	143	

SOURCE: Computed from records of ONERN.

TABLE 17

CLIMATIC DATA, VITOR, LA JOYA ANTIGUA
1934-1935 THROUGH 1967-1968

Years	Temperature		Precipitation (mm)				
	Coldest Month Average (°C)	Warmest Month Average (°C)	Oct. Dec.	Jan. Mar.	Apr. June	July Sept.	Total
1934-1935	15.7	18.8	a	a	a	a	a
1935-1936	b	b	19.0	13.8	0.0	0.0	33.6
1936-1937	b	b	3.1	13.2	0.0	0.0	16.3
1937-1938	15.9	19.6	0.0	100.1	0.0	0.0	100.1
1938-1939	17.3	18.7	0.0	34.3	0.0	0.0	34.3
1939-1940	17.3	18.7	0.0	4.1	0.4	1.6	6.1
1940-1941	15.3	20.1	0.0	9.4	5.3	0.6	15.3
1941-1942	13.4	21.3	3.5	17.8	0.0	0.0	21.3
1942-1943	17.3	20.3	5.6	30.6	2.2	0.7	39.1
1943-1944	16.1	18.0	3.8	74.0	6.8	1.0	85.6
1944-1945	16.6	19.3	1.8	9.1	0.0	2.2	13.1
1945-1946	17.9	20.3	3.4	25.0	2.2	0.0	30.6
1946-1947	b	b	2.9	1.9	0.0	1.2	6.0
1947-1948	14.2	20.4	0.6	11.0	3.8	0.0	15.4
1948-1949	b	b	2.0	14.8	4.4	0.0	21.2
1949-1950	b	b	1.5	1.8	0.0	1.5	4.8
1950-1951	15.4	19.6	4.5	7.3	0.0	1.1	12.9
1951-1952	14.4	19.6	0.5	11.7	0.4	0.6	13.2

TABLE 17--Continued

1952-1953	16.2	18.8	0.6	7.9	0.4	0.5	9.4
1953-1954	15.8	20.0	0.4	14.4	1.4	0.0	16.7
1954-1955	15.4	14.2	1.2	32.6	0.4	0.0	34.2
1955-1956	19.0	15.6	1.2	2.1	0.0	0.0	3.3
1956-1957	15.4	18.8	0.0	1.0	0.0	0.0	1.0
1957-1958	b	b	b	1.9	0.0	0.0	b
1958-1959	16.8	19.1	0.0	23.3	0.9	0.0	24.2
1959-1960	b	b	b	2.0	0.0	0.0	b
1960-1961	16.8	19.2	0.0	21.0	0.0	0.0	21.0
1961-1962	15.6	18.2	1.0	1.6	0.0	0.0	2.6
1962-1963	16.8	19.9	0.0	12.1	0.0	4.0	16.1
1963-1964	16.5	20.5	0.0	4.6	2.5	0.0	7.1
1964-1965	b	b	b	0.0	0.0	3.5	b
1965-1966	15.0	19.0	0.0	2.4	6.0	0.0	8.4
1966-1967	15.2	20.0	3.0	69.3	0.0	0.0	72.3
1967-1968	15.4	18.8	0.0	85.8	2.5	0.0	88.3
1968-1969	b	b	0.0	0.0	b	b	b
Yearly Average	16.0	19.4	1.9	19.5	1.2	0.6	25.8
Yearly Median			0.6	11.4	0.0	0.0	16.2

SOURCE: Computed from records of ONERN.

a Precipitation data lacking for 1934-1935

b Part of year lacking.

TABLE 18

CLIMATIC DATA, AREQUIPA CITY, (CORPAC)
1949-1950 THROUGH 1970-1971

Years	Temperature		Precipitation (mm)				
	Coldest Month Average (°C)	Warmest Month Average (°C)	Oct. Dec.	Jan. Mar.	Apr. June	July Sept.	Total
1949-1950	13.3	16.1	4.9	26.0	0.0	0.0	30.9
1950-1951	14.1	16.7	18.9	66.2	0.0	0.0	85.1
1951-1952	13.2	17.4	5.7	143.2	1.0	0.0	149.9
1952-1953	13.6	17.2	6.9	242.0	0.0	0.5	249.4
1953-1954	12.7	17.4	5.2	73.6	0.0	0.0	78.8
1954-1955	12.5	16.9	7.3	462.4	0.0	0.0	469.7
1955-1956	12.7	16.8	3.3	48.2	0.0	0.0	51.5
1956-1957	13.4	16.1	0.0	49.1	0.0	2.6	51.7
1957-1958	13.1	17.3	2.8	55.4	0.0	0.0	58.2
1958-1959	13.7	16.9	0.0	129.0	0.4	0.0	129.4
1959-1960	12.5	16.5	0.0	17.4	0.0	0.0	17.4

TABLE 18--Continued

1960-1961	14.9	16.7	4.0	159.0	0.0	0.0	163.0
1961-1962	14.4	17.5	20.6	46.2	0.0	0.0	66.8
1962-1963	13.9	17.6	0.0	178.8	1.0	5.0	184.8
1963-1964	14.6	18.0	0.0	13.9	2.0	0.0	15.9
1964-1965	15.3	17.4	27.3	24.4	0.0	9.0	60.7
1965-1966	14.7	18.0	0.0	6.5	3.3	0.0	9.6
1966-1967	14.0	17.2	3.4	174.3	7.5	0.0	185.2
1967-1968	14.2	18.0	0.0	72.5	0.0	0.0	72.5
1968-1969	16.0	18.0	0.0	16.9	5.8	0.0	22.7
1969-1970	15.4	18.3	7.0	46.6	0.0	0.0	53.6
1970-1971	14.7	17.6	0.0	12.8	0.2	0.0	13.0
Yearly Average	14.0	17.3	5.3	93.8	1.0	0.8	100.9
Yearly Median	14.0	17.4	3.4	52.3	0.0	0.0	63.8

SOURCE: Computed from records of ONERN.

TABLE 19

CLIMATIC DATA, IMATA
1936-1937 THROUGH 1970-1971

Years	Temperature		Precipitation (mm)				
	Coldest Month Average (°C)	Warmest Month Average (°C)	Oct. Dec.	Jan. March	April June	July Sept.	Total
1936-1937	a	a	224.1	767.0	28.7	29.3	1049.1
1937-1938	a	a	a	504.5	132.5	8.9	a
1938-1939	2.8	6.6	48.2	366.2	45.6	43.0	503.0
1939-1940	0.6	8.0	68.6	180.1	88.1	2.3	339.1
1940-1941	a	a	75.1	236.0	75.6	13.9	400.6
1949-1942	a	a	150.0	211.2	28.5	3.2	392.9
1942-1943	0.1	5.2	45.7	419.8	51.5	5.1	522.1
1943-1944	1.4	4.3	78.2	373.9	98.6	7.2	557.9
1944-1945	0.4	5.5	135.1	330.5	7.3	51.7	524.6
1945-1946	0.6	4.4	94.5	342.8	87.0	48.3	572.6
1946-1947	-0.4	5.8	301.0	492.2	84.8	24.1	902.1
1947-1948	-0.2	7.3	97.8	464.4	131.0	27.6	720.8
1948-1949	0.4	5.6	154.6	554.4	64.4	62.2	835.6
1949-1950	-2.9	5.9	169.3	311.7	85.2	8.6	574.8
1950-1951	a	a	158.6	438.0	a	a	a
1951-1952	a	a	a	294.7	17.1	30.4	a
1952-1953	-1.6	5.6	72.5	415.4	29.7	21.9	539.5
1953-1954	-0.6	5.8	202.1	424.7	73.1	8.5	708.4

TABLE 19--Continued

1954-1955	a	a	a	484.1	41.9	7.6	a
1955-1956	-2.4	5.2	149.3	244.7	0.0	41.0	435.0
1956-1957	-0.6	5.9	50.9	260.6	33.0	8.4	352.9
1957-1958	-0.4	6.4	89.7	325.4	4.4	3.1	422.6
1958-1959	0.0	5.6	96.0	312.9	30.3	0.4	439.6
1959-1960	-0.8	5.6	164.3	286.5	26.3	29.0	506.1
1960-1961	-0.4	6.0	152.8	440.5	74.5	42.8	710.6
1961-1962	-1.6	5.6	204.9	281.2	77.5	23.0	586.6
1962-1963	-0.8	4.8	75.5	397.6	41.1	11.8	526.0
1963-1964	-1.6	6.2	187.0	345.5	69.2	7.0	608.7
1964-1965	-2.2	4.3	70.9	215.9	13.7	4.1	304.6
1965-1966	-1.4	5.2	88.8	231.6	51.4	3.0	374.8
1966-1967	-1.0	4.9	169.1	69.4	23.1	16.5	278.1
1967-1968	-1.4	4.6	38.5	407.3	18.3	6.4	470.5
1968-1969	-1.2	5.1	211.6	268.6	19.9	13.7	513.8
1969-1970	-1.4	5.0	144.8	359.0	32.2	5.0	541.0
1970-1971	-2.8	4.4	60.4	389.7	17.8	0.0	467.9
Yearly Average	-0.69	5.53	125.9	355.7	50.1	18.2	538.1
Yearly Median			116.5	345.5	41.5	10.4	522.1

SOURCE: Computed from records of ONERN.

[a]Part of year lacking.

TABLE 20

LENGTH OF PERIODS OF CONTINUOUS DROUGHT

Station	Frequencies												
	Number of Continuously Dry Months												
	2	3	4	5	6	7	8	9	10	11	12	13	Total
Vitor	a	a	6	6	4	1	1	7	2	1	0	1	29
Arequipa (corpac)	a	a	3	4	2	5	3	4	2	0	0	0	23
Imata	11	3	2	0	0	0	0	0	0	0	0	0	16

SOURCE: Computed from records of ONERN.

[a]Not tabulated.

is greater on the Pampa of La Joya than in the Campiña. The longest drought in the Campiña has been 10 months. This has happened twice, once from March to December 1956, and once from April 1958 to January 1959. There have been four periods of continuous drought for ten months or longer on the Pampa of La Joya, the longest lasting for 13 months from May 1968 to May 1969. With a decrease in variability of precipitation, the upper basin experiences much shorter periods of continuous drought: the longest recorded at the Imata station was four months. This occurred twice, once from May to August 1953 and again from April to July 1956.

Potential Evaporation

Another manner of analyzing the climate is by means of the water balance. Two stations in the study area, Vitor on the Pampa of La Joya and La Pampilla in the Campiña, provide data for water balance computations which use potential evapotranspiration rates as calculated by the Peruvian National Office for the Evaluation of Natural Resources (ONERN).[1] After evaluating several methods for

[1]ONERN, Inventario, Evaluación y Uso Racional de los Recursos Naturales de la Costa: Cuencas de los Rios Quilca y Tambo, I, 524-525 and III, 201-202.

determining potential evapotranspiration, ONERN used the following empirical formula developed by C.J. Grassi and J.E. Christiansen:[1]

$$Etp = (5.46\ CR)\ (CcLc)\ (CT)\ (CTD)$$

where:

CR = 0.1824 + 0.0575R; where R is the theoretical solar radiation at the top of the atmosphere in mm/day

CcLc = 1.15 - 0.08 cLc; where cLc is cloudiness expressed in tenths

CT = $0.62081 + 0.02633T - 0.0003682T^2$; where T is the mean temperature in C°

CTD = 0.9361 = 0.00767TD; where TD is the difference between the maximum and mean temperature

According to ONERN, this formula results in the best approximation of reality. This formula also overcomes the problem of accounting for the elevated height of the Chili Basin.

In the study area potential evapotranspiration, influenced by solar radiation, results in a greater water need during spring than at other times of the year (figures 5 and 6). Since precipitation varies markedly throughout the year and the small amounts that do occur come after the high water needs in the spring, there is a decided deficit during every months at both stations (table 21). Therefore, irrigation is required during every month for plant growth.

On comparing the climate for the years that this study covers with the period of record, the mid-1960s were found to be generally dry, although 1966 registered precipitation well over the average (tables 17 and 18). Temperatures, however, vary much less than precipitation from year to year.

Surface Water

Most water used in the Chili Basin is from surface supplies. The Chili River, which flows through the study area, receives the bulk of its water supply from the high sierra of the upper basin of the Quilca watershed. In recent years modifications have taken place in the natural stream flow; structural adjustments smooth out seasonal and annual fluctuations, producing a larger useable

---

[1] For an evaluation of an additional 24 equations for estimating evapotranspiration in Peru see, Fernando Chanduví-Acuña, "A Study of Evaporation and Evapotranspiration in Peru." Master of Science thesis, Utah State University, Logan Utah, 1969.

Fig. 5

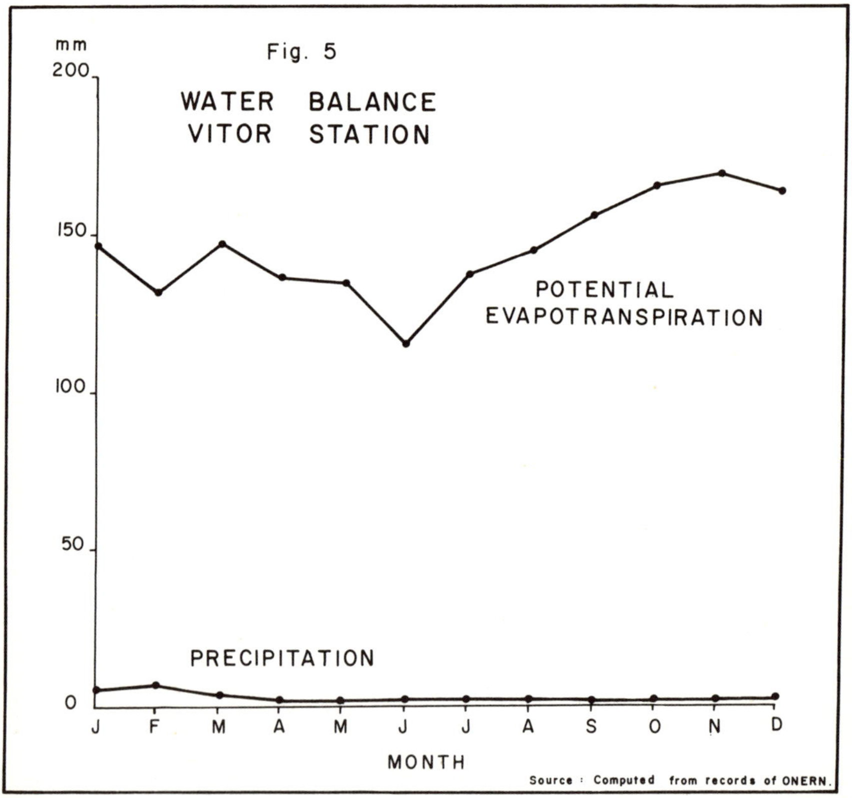
mm
200
150
100
50
0
WATER BALANCE
VITOR STATION
POTENTIAL
EVAPOTRANSPIRATION
PRECIPITATION
J
F
M
A
M
J
J
A
S
O
N
D
MONTH
Source : Computed from records of ONERN.

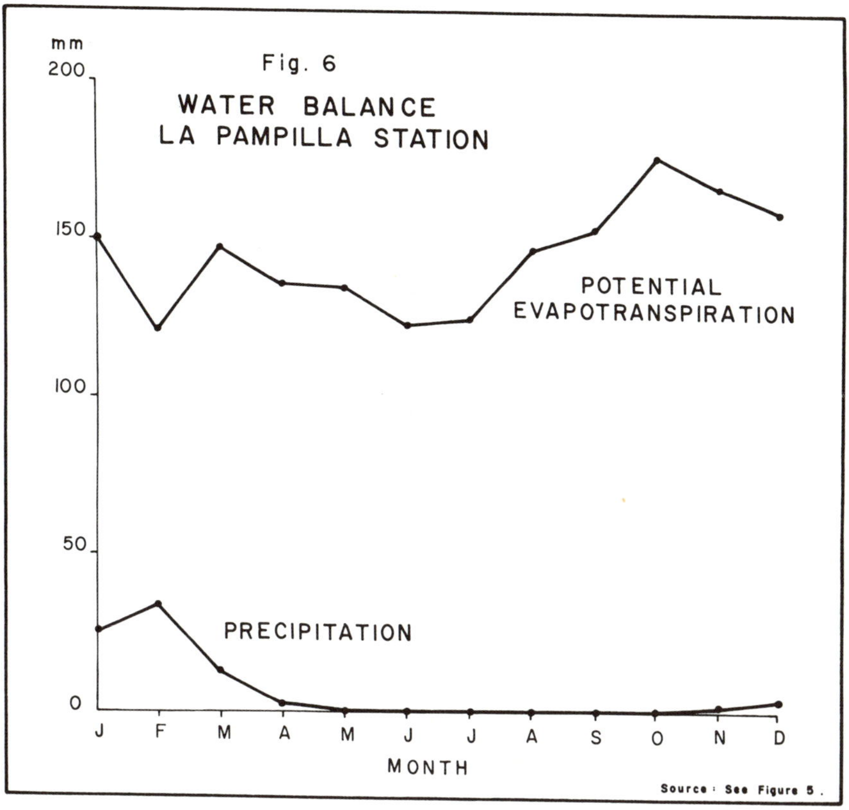
mm
200
150
100
50
0
Fig. 6
WATER BALANCE
LA PAMPILLA STATION
POTENTIAL
EVAPOTRANSPIRATION
PRECIPITATION
J F M A M J J A S O N D
MONTH
Source: See Figure 5.

TABLE 21

MOISTURE DEFICIT IN CHILI BASIN

Month	Moisture Deficit (mm) Average Climatic Data	
	Vitor	La Pampilla
January	142	123
February	122	91
March	146	136
April	138	135
May	137	133
June	117	122
July	139	124
August	146	146
September	157	151
October	165	173
November	167	163
December	162	155
Total	1,738	1,652

SOURCE: Computed from records of ONERN.

supply. The allocation of the water follows both longstanding traditional practices and more modern innovative procedures. The surface water is assessed in terms of the natural flow, human modifications of flow, and the methods of water allocation.

Source of Data

Within the Quilca Basin there are several gaging stations (figure 7). The oldest station, Charcani, established in 1923 at the head of the Chili Basin, was located in the intake canal just above the sandtrap for the hydroelectric station (Charcani III) and about 11 km upstream from the city of Arequipa. In 1962, the government moved the station upstream about 1,000 m, near the entrance to the intake canal of the new hydroelectric station (Charcani IV). The data from the Charcani gaging station are reliable, since here is a fairly permanent cross-section, and conditions are ideal for making proper measurements by automatic

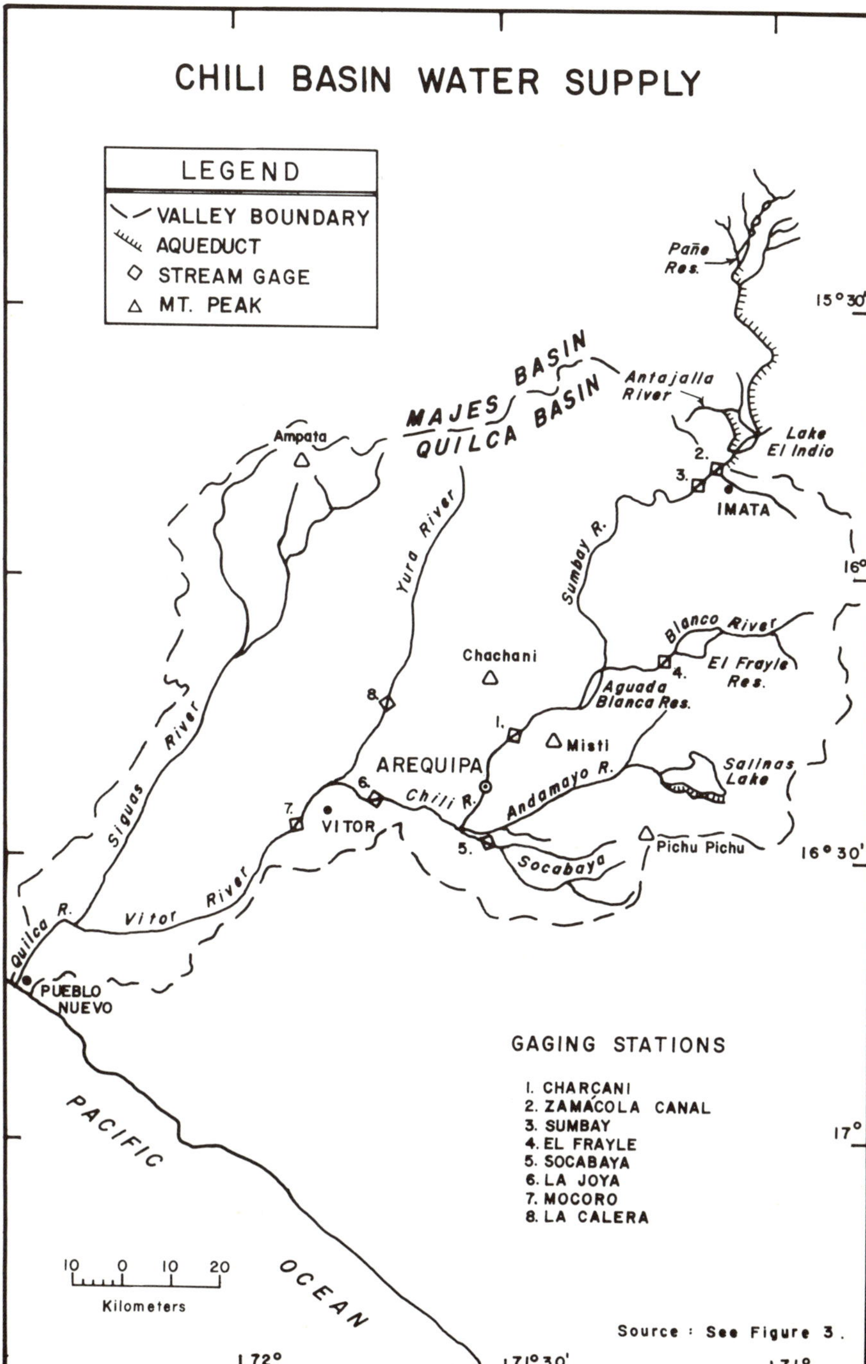

Fig. 7 - Chili Basin Water Supply.

gaging. The float method is used during floods.[1] Since the collection of data from this station stopped from 1930 to 1937, the record from 1937 onward is used. At an elevation of 2,752 m, this station controls 4,145 km^2 of the humid basin or 55 percent of the total humid Quilca watershed. The government aptly located this station, since there are no major diversions upstream from this point.

Three additional gaging stations, installed upstream from the Charcani station, monitor flows that have recently undergone modifications. One of these stations located in the Zamácola Canal about 1 km upstream at a point where it empties into the Sumbay River was established in 1945. Technicians, using a current meter, measure the daily flow. While the data seem to be fairly reliable, ONERN believes the site should be changed in order to solve the problem of sedimentation.[2] About 200 m downstream from where the Zamácola Canal empties into the Sumbay River is another gaging station established a year earlier, in 1944.

The third gaging station upstream from the Charcani station is located on the Blanco River, approximately 50 m downstream from El Frayle Dam. Although the station was installed in 1953, meaningful data were not available until 1964. There is also a problem with the reliability of the data obtained there, due to the influence of the backwater from the Cazadones Ditch that empties into the Blanco River a few meters downstream from the gaging station.[3]

Two additional gaging stations are located downstream from the Charcani station, one near the mouth of the Socabaya River and the other at the entrance to the La Joya Irrigation Project. Both of these stations were established at about the time that this study was completed. In past years, two other gaging stations did exist but were only operated for about five years. One was on the Vitor River near Mocoro and the other on the Yura River at La Calera. The reliability of the data from these two stations, however, is questionable.[4]

From among this set of gaging stations, three stand out as critical in assessing surface water supplies. One important set of stream flow data analyzed is from the strategically located Charcani gaging station. The date reflect stream flow conditions in the upper basin just before this water is put to use by man for irrigation purposes. Data from the Zamácola Canal and El Frayle gaging stations

---

[1]ONERN, Inventario, Evaluación y Uso Racional de los Recursos Naturales de la Costa: Cuencas de los Rios Quilca y Tambo, 1:414.

[2]Ibid., p. 413. [3]Ibid., p. 414. [4]Ibid., p. 411.

aid in assessing interbasin transfers and inbasin storage, respectively.

The water year used begins in October, which generally corresponds with the end of the dry period, and ends in September the following year. This is the water year used by the Peruvian hydrologic agencies.

Natural Flow

The principal source of irrigation water for the Campiña and the Pampa of La Joya is the Chili River, which is formed by the confluence of the Blanco and the Sumbay Rivers (figure 7). The origin of the Quilca River is in the Andes, at an elevation well above 3,000 m. Even though there are snow fields and glaciers in the upper basin of the Quilca watershed, melt water does not greatly influence the average monthly flow.[1] However, variations in climatic conditions, such as timing and duration of precipitation in the upper basin of the Quilca, do influence the hydrologic regime in the Chili Basin. Since there are well-defined wet and dry periods in the water supply area, there is marked seasonality in the flows of the Chili River. In like manner, variations in annual precipitation in the upper basin result in fluctuations in annual flow downstream in the Chili Basin.

The average annual flow of the Chili River at the Charcani gaging station is approximately 12 m^3/s (table 22). The annual flow ranges from 5 m^3/s to 36 m^3/s. In over half of the years, a flow of less than the overall average occurred. In fact, during 6 years (1940, 1950, 1959, and 1964-66) the Chili River had a discharge of less than 50 percent of the average annual flow. Yet there were 9 years when the discharge was over 140 percent of the average annual flow (table 23).

The Quilca River is one of the few in southern Peru that flows year-round into the Pacific Ocean. The coefficient of irregularity (Cr) for the Chili River as measured at the Charcani gaging station is 0.38 which is somewhat lower than most of the coastal rivers of Peru. However, this variation is more than that of two other important rivers in the immediate locality, the Majes and the Tambo Rivers, with Cr of 0.29 and 0.34, respectively.[2]

In general, these are two distinct seasonal stream flow periods

---

[1] United Nations, Economic Commission for Latin America, "Los Recursos Hidráulicos de America Latina: Perú," p. 59.

[2] Ibid., p. 64.

TABLE 22

NATURAL FLOW OF CHILI RIVER AT CHARCANI
1937-1938 THROUGH 1970-1971

Month	Average Flow		Median Flows		
	m^3/seg.	Percent of Total	Median Low m^3/s	Median Flow m^3/s	Median High Flow m^3/s
October	3.60	2.5	1.20	3.59	4.46
November	3.78	2.7	1.61	3.85	6.10
December	6.23	4.5	1.94	4.68	32.90
January	21.02	15.3	2.91	13.03	115.26
February	35.58	25.9	4.26	24.70	160.80
March	32.63	23.7	4.06	19.30	106.71
April	13.35	9.7	3.18	6.73	56.00
May	5.35	3.9	2.69	4.37	22.33
June	4.15	3.0	2.64	4.07	5.50
July	4.05	2.9	2.59	4.12	5.15
August	3.97	2.9	2.63	3.94	5.35
September	3.84	2.8	1.08	3.95	5.49
Year	137.50	99.9	30.79	96.33	526.05

SOURCE: Computed from records of ONERN.

during a water year (table 24). The one occurring during the first three months of the year accounts for 65 percent of the total annual stream flow. The remaining flow is about evenly divided among the other quarters of the year.

Modifications of Stream Flow

Until 1945, the natural flow of the Quilca River served the needs of farmers fairly well, but from then until the present a series of major modification in the flow have taken place in order to meet the growing needs brought about by the development of new irrigated areas by government and private groups (figure 7). Even

TABLE 23

DEVIATIONS FROM NATURAL FLOW OF THE CHILI RIVER AT CHARCANI 1937-1938 THROUGH 1970-1971

Month	Number of Years with Flows that were Specified Percentage of Average				
	<60%	61%-79%	80%-120%	121%-140%	>140%
October	1	1	29	3	--
November	1	3	28	1	1
December	2	21	7	2	2
January	17	5	4	1	7
February	16	5	4	--	9
March	17	3	3	--	11
April	19	4	3	1	7
May	2	12	17	--	3
June	--	4	25	5	--
July	--	6	23	5	--
August	--	8	22	4	--
September	1	2	27	3	1
Year	11	7	6	1	9

[1] SOURCE: Computed from records of ONERN.

with modifications, the flow of the Quilca River largely originates within the basin itself. There are, however, interbasin transfers from the adjacent Majes River to the Quilca watershed. Three reservoirs, which have a combined capacity of 340 million m^3, have been constructed to augment and regulate the flow of the Quilca River. According to ONERN, the present modified flow adequately covers the demands for water in the Basin. However, if there is an increase in the area irrigated, there will be a need for further modifications.[1]

[1] ONERN, Inventario, Evaluación y Uso Racional de los Recursos Naturales de la Costa: Cuencas de los Rios Quilca y Tambo I, 409.

TABLE 24

AVERAGE NATURAL FLOW OF CHILI RIVER AT CHARCANI
1937-1938 THROUGH 1970-1971

Natural Flow							
Oct. - Dec.		Jan. - Mar.		Apr. - June		July - Sept.	
m^3/s	% of Total Flow	m^3/s	% of Total Flow	m^3/s	% of Total Flow	m^3/s	% of Total Flow
13.6	10	89.2	65	22.9	17	11.9	8

SOURCE: Computed from records of ONERN.

The first interbasin transfer took place in 1945. Water from Lake Jauyulaca (El Indio) in the Majes watershed was diverted to the Sumbay River in the Quilca Basin by way of the Zamácola Canal, a length of 7.5 km. Due to the large amount of infiltration and evaporation (2.5 m^3/s) along this unlined canal, less than 1 m^3/s reaches the lands intended for irrigation, the project of Zamácola in the Campiña.

Another interbasin transfer, the largest and most important, is the Pañe Reservoir complex, which started functioning in 1965.[1] Water impounded in the Upper Colca River, a tributary of the Majes River, flows to the Upper Sumbay River largely by means of a series of canals (figure 7). This interbasin transfer begins at the Pañe Reservoir and follows the Pañe River a few hundred meters, branching off to connect with the Bamputane River, still in the Majes Basin, by way of the Pañe-Bamputane Canal, 5,162 m long with a capacity of 6 m^3/s.[2] The next stage, the largest of this interbasin transfer, the Bamputane-Sumbay Canal, actually allows the water to cross the divide into the Upper Quilca. This canal, 53.2 km long with a capacity that varies from 6 to 16 m^3/s, follows the Bamputane River for about 5 km. It then branches off to the south to traverse open country and cross three principal rivers: the Negrillo, the Blanquillo, and the Colca. The canal traverses the first two rivers by siphons and the latter by a covered conduit. From here water reaches Lake

[1] Ibid., 1:410.

[2] Ibid., 1:415.

El Indio through a 2,000 m long canal. Upon leaving Lake El Indio, the route follows the original Zamácola Canal of 1945, which has been enlarged. An addition to this complex is a canal 12 km in length that taps the Antasalla River in mid-reach and carries the flow south to southeast where it empties into the original Zamácola Canal near the outlet of Lake El Indio.

The Pañe Reservoir complex furnishes water for the expansion of the La Joya Irrigation Project. The amount of water to be transferred to the Quilca River from this project amounts to about 4.4 m^3/s. A large amount, however, is lost through infiltration and evaporation resulting in a diminished supply of 1 to 2 m^3/s.[1] Nevertheless, diverting water from the Pañe Reservoir to the Quilca Basin has markedly increased interbasin transfers (table 25).

TABLE 25

INTERBASIN TRANSFERS, QUILCA BASIN
1945-1946 THROUGH 1970-1971

Additions	Average Annual Flow m^3/s
Zamácola Canal	
1945-1946 through 1964-1965	0.67
1965-1966 through 1970-1971	3.44

SOURCE: Computed from records of ONERN.

The third interbasin transfer is of only minor importance in supplying water to the Quilca Basin. Spring water from the closed basin dominated by Lake Salinas is transferred to the headwaters of the Andamayo River, a tributary of the Socabaya River (figure 7). This interbasin transfer has a flow of less than 1 m^3/s and supplies irrigation water to areas along the Andamayo River.

Of the three principal reservoirs affecting the stream flow in the study area, El Frayle is the largest, with a design capacity of 200 million m^3 (figure 7). El Frayle is located on the Blanco River, in the lower end of a canyon just below the point where the Pasto Grande River enters the Blanco. This reservoir was begun in 1956, but lay idle for several years after its completion in 1959. In 1966,

[1] Ibid.

a rupture developed in an abutment, postponing its full use. Because of its recent construction and the interruption caused by structural failure, the precise operational characteristics of this reservoir had not been determined at the time of this study.

The other reservoir in the Upper Quilca Basin, the Aguada Blanca, is located in the canyon of the upper Chili River (figure 7). It extends upstream almost to the point where the Blanco River and the Sumbay River form the Chili River. The dam, completed in 1971, impounds water from the Sumbay River. With a design capacity of 40 million m^3, the reservoir is not of sufficient size to capture all the runoff. Government officials have suggested that an additional reservoir be built on the El Pillones River to capture the excess flows of the Sumbay.[1] Apart from supplying water to the expanded La Joya Project, the Aguada Blanca Dam will provide the necessary flow for a new addition to the hydroelectric plant on the Chili River (Charcani V) with a capacity of 60 MW.[2]

The third reservoir modifying the water supply of the Chili River is outside the Quilca Basin on the Pañe River within the Upper Colca watershed. Water stored in the Pãne reservoir is transferred to the Quilca River by means of the Pañe-Zamácola Canal complex, as noted above. The design capacity of this reservoir is 85 million m^3.

The Department of Irrigation of the Ministry of Public Works manages all three dams. The planned operation of these dams calls for managing them in tandem.[3] The Pañe Dam remains closed for the first four months of the year, storing water in its reservoir during the rainy season. Meanwhile, the Aguada Blanca Dam opens and delivers water by way of the Chili River to the downstream irrigation projects until all water drains out of its reservoir. Then in the fifth month of the year, the Pañe Dam opens to release water. After El Pañe is drained, downstream water needs are supplied by releasing water from El Frayle Dam. With the advent of the rainy season in December or January, the cycle of operation is repeated. In this fashion, according to ONERN, the proposed operation of the series of dams and interbasin transfers from the Upper Colca can give a

---

[1]Ibid., 2:572.

[2]United Nations, Economic Commission for Latin America, "Los Recursos Hidráulicos de America Latina, Perú," p. 170.

[3]ONERN, Inventario, Evaluación y Uso Racional de los Recursos Naturales de la Costa: Cuencas de los Rios Quilca y Tambo 1:420.

planned regulated flow of 13 m^3/s, with the assurance of covering all needs up to 70 percent of the time.

As a result of these modifications, the actual flow into the Chili Basin is now greater than the natural flow (table 26). Also, these modifications make for less peaking during the beginning of the year. This is especially noticeable after 1965, when the average quarterly flow from July through September increased from 11 percent to 22 percent of the average annual flow.

Water Allocation

Until the end of the 1960s, the Water Code (Código de Aguas) of 1902 was the basic legal document governing the allocation of water in Peru.[1] On June 24, 1969, the General Water Law (Ley General de Aguas, Decreto Ley no. 17752), was promulgated, and a sweeping change occurred in the administration of the country's water resources. Until this time, the allocation of water was not always in harmony with the public interest.

Although through the years continuing change has taken place in the manner of allocating irrigation water, there is an ever-present link to past procedures that can be traced from Inca times through the Colonial and Republican periods to the present. The legacy of the Incas still exists.

The overthrow of the Incas by the Spanish forces of Pizarro in 1532 created a chaotic water management situation.[2] To ameliorate this condition, Charles V of Spain in 1536 issues from Valladolid a royal decree authorizing the continuation of the Inca system of water allocation.[3] Following this edict, however, the government paid little attention to the equitable allocation of water among farmers.

In time, through the powers granted by the Council of the Indies, the Viceroys of Peru prepared several legal documents affecting judgments on water disputes. In 1556, one of these early documents created special judges of water (juzgados privativos de aguas), intending to assure more order and vigilance in the distribution and management of irrigation water. Some 20 years later,

---

[1] Carlos Rodriguez Escobedo, Código de Aguas, Concordancias de Legislación y Jurisprudencia: Leyes y Reglamentos de Irrigación (Arequipa, Perú: Editorial El Deber, S.A. Ltda., 1956).

[2] Juan Vicente Nicolini, "La Policía de las Aguas en el Perú," Revista Universitaria de la Universidad Mayor de San Marcos (Lima, Perú) 15, no. 2 (1920), p. 246.

[3] "Las Administraciones Técnicas de Aguas de Regadío y la Necesidad de su Subsistencia," Aguas e Irrigación (Lima, Perú) 1, no. 1 (May 15, 1932), p. 83.

TABLE 26

ACTUAL AND NATURAL INFLOW
CHILI BASIN AT CHARCANI

Flow (m /s)	Actual Inflow	Natural Inflow	Flow (m /s)	Actual Inflow	Natural Inflow
1945-46 Through 1964-65			1965-66 Through 1970-71		
Average annual inflow	50.80	47.76	Average annual inflow	42.52	34.32
Percent of natural	106.37	100.00	Percent of natural	123.89	100.00
Average Oct.-Dec. flow	5.71	4.86	Average Oct.-Dec. flow	8.26	5.76
Percent of average annual	11.24	10.18	Percent of average annual	19.43	16.78
Percent of natural	117.49	100.00	Percent of natural	143.40	100.00
Average Jan.-March flow	32.68	31.89	Average Jan.-March flow	15.58	19.20
Percent of average annual	64.33	66.77	Percent of average annual	36.64	55.94
Percent of natural	102.48	100.00	Percent of natural	81.15	100.00
Average Apr.-June flow	8.18	7.57	Average Apr.-June flow	9.62	5.71
Percent of average annual	16.10	15.85	Percent of average annual	22.62	16.64
Percent of natural	108.00	100.00	Percent of natural	168.48	100.00
Average July-Sept. flow	4.63	3.82	Average July-Sept. flow	9.17	3.85
Percent of average annual	9.11	8.00	Percent of average annual	21.54	11.22
Percent of natural	121.20	100.00	Percent of natural	238.18	100.00

SOURCE: Computed from records of ONERN.

on an extensive trip through the country, Don Francisco de Toledo, viceroy of Peru, observed a haphazard and negligent control in the distribution of water, even with the appointment of these special water judges. The water ordinance issued by Don Francisco the following year was the most important decree formulated by any viceroy of Peru on this subject.[1] In this ordinance of 16 articles, many measures were specified for the proper distribution and management of water, as well as a new procedure for appointing special water judges.

Not until the last century of the Colonial period did authorities begin to develop laws on the allocation of irrigation water. Several regulations drawn up systematized the allocation of water only for one or two regions, nothing applied to the entire country. Two of these regulations were more lasting and influential than others. Ambrosio Cerdán y Pontero in 1785 developed water legislation for the Rimac Valley, the location of the capital, Lima, and Deán Saavedra produced a regulation for several valleys in the Department of La Libertad in 1699.[2] The original text of Cerdán, entitled "A General Work on the Waters that Enrich the Valleys of Lima," was according to Nicolini simply a good study of water legislation, containing many important scientific and historical principles. When Dr. Francisco García Calderón reworded Cerdán's material, however, he developed a truly practical and useful set of water regulations for the Rimac Valley.[3] At the close of the Colonial period there was still no one single declaration of water allocation for the entire country; in fact, many valleys had no formally written regulation for the distribution of water. This piece-meal approach in setting standards of water allocation throughout Peru persisted for many years.

Even with the formation of the Republic of Peru, the management of irrigation water continued in disarray. In many instances the municipalities took the initiative in managing irrigation water, thereby worsening an already critical situation.[4] Finally the

---

[1] Nicolini, "La Policía de las Aguas en el Perú," p. 248.

[2] Carlos Malpica, "El Régimen Legal de las Aguas," Revista de Economía y Agricultura (Lima, Perú) 1 (June-August 1964):310.

[3] Nicolini, "La Policía de las Aguas en el Perú," pp. 249-250.

[4] "Las Administraciones Técnicas de Aguas de Regadío y la Necesidad de su Subsistencia," p. 83.

government realized the futility of such an uncoordinated water management policy and began dictating a series of resolutions regulating the allocation of water. In 1836, by Supreme Resolution, the government reestablished the special water judgeships comparable to those of the Colonial period. Five years later, twenty years after the founding of the republic, one of the first major supreme decrees extended the Cerdán Ordinance throughout coastal Peru.[1] Many other laws appeared but all were inefficient since they did not result in one water code for the entire country.[2] The first of several attempts at codifying the water law of Peru began in 1870. In 1899, the last commission was appointed. This commission finally developed a unified national water code, which was passed by congress and signed into law by the president in 1902.

The Water Code of 1902, reflecting in large part the Spanish water law of 1876, did not take into consideration the unique milieu of Peru. This resulted in an inadequate law.[3] While all rivers were under the public domain, water rights existing before the enactment of this code remained intact.[4] Such practices as toma libre, which allowed a diverter as much water as he wished, and usos y costumbres, a practice permitting unique allocation procedures for an area, continued to exist. This 1902 code also did not modify water rights among farmers using the same canal.[5] All of these practices were in vogue in the Arequipa study area during this survey.

In the mid-1960s, President Belaunde instituted a new agrarian reform program, which included changes in the administration of water throughout the country. While this new reform did not affect many of the old water rights, it did abolish the custom of cabecera, which allowed farmers at the head of a canal daily access to water, and farmers farther downstream less frequent access.[6] This is the

---

[1]Helen L. Clagett, A Guide to the Law and Legal Literature of Peru, Latin American Series no. 20 (Washington, D.C.: Government Printing Office, 1947), p. 125.

[2]"Las Administraciones Técnicas de Aguas de Regadío y la Necesidad de su Subsistencia," p. 83.

[3]Ibid., p. 84, and "Código de Aguas: Anacrónico; Causa Injusticia en Reparto," La Prensa (Lima, Perú), 1 January 1963, p. 7.

[4]Malpica, "El Régimen Legal de las Aguas," p. 312.

[5]E. Lasarte, "Estudio Relativo a la Reglamentación de los Derechos de Agua de Regadío," Aguas e Irrigación (Lima, Perú) 1, no. 2 (June 1932), pp. 12-13.

[6]Malpica, "El Régimen Legal de las Aguas," p. 312.

case in Arequipa. In addition, the new law perpetuated the existing profusion of uncoordinated public water agencies. For instance, La Dirección de Aguas e Irrigación, Ministerio de Fomento y Obros Públicas (D.A.I.), while charged with constructing the infrastructure and managing water allocation from the major reservoirs, was not fully coordinated with La Dirección de Aguas de Regadío, Ministerio de Agricultura (D.A.R.), which was charged with allocating water directly to diverters.[1]

In 1969, a radical change took place in the allocation of water throughout the country. The General Water Law annuls all previous laws pertaining to water, including the 1902 Water Code, and all other vested rights in water and institutes a new order in water distribution and rights based on scientific principles reflecting a concern for the social welfare of all the people of Peru.[2]

Some irrigated regions in Peru were spared gross inequities in allocating water to diverters. Such is the case in the Chili Basin, where there are few large and powerful landowners. Nevertheless, the complex system in the basin requires constant vigilance and maintenance in the allocation of water.

For several years previous to the initiation of this study, and continuing during most of the time field work was in progress, the D.A.R. administered the allocation of irrigation water. This agency has many regional offices throughout the irrigated areas of the country. The principal office for the Chili Basin is in the city of Arequipa, with a subbranch in the town of Vitor to allocate water on the Pampa of La Joya. The principal office in Lima appoints the head of the regional D.A.R., the administrator-engineer. Among the more important duties of this office are the determination of the available water supply, its allocation, and its supervision among diverters.

Under the aegis of the D.A.R., various councils of irrigators, juntas de regantes, assist in the administration of water. Normally each major irrigation ditch has its own junta de regantes of 5 farmers from that area, which helps control the allocation of water.

An observation made in the 1960s by a leading United States

---

[1]Guillermo J. Cano, Gobierno y Administración de los Recursos Hídricos del Perú (Lima: Oficina Nacional de Evaluación de Recursos Naturales, 1966), p. 6.

[2]República del Perú, Ley General de Aguas, Decreto Ley No. 17752, 24 de Julio de 1969 (Lima: Empresa Editora del Diario Oficial "El Peruano," 1969). Also see Marcelino Gavilan, Análisis y Características de la Nueva Ley General de Aguas del Perú (Rome: Food and Agricultural Organization, September 5, 1969) and "Peru's Water Reform Law," Peruvian Times, April 10, 1970, pp. 11-13.

authority on irrigation, about water allocation in Peru, is applicable to the Chili Basin. After having made an extensive survey of the water resources of coastal Peru, and in discussing the problems involved in measuring river flows and water distribution, Christiansen remarked that, in general:

> Streamflow of each river is theoretically divided among the many canals diverting water from the stream in accordance with water rights, which appear to be somewhat complicated but are essentially based on priority of use and custom (*usos y costumbres*). For each river there is usually a tabulation of all of the rights under each canal, giving the amount of flow or the percentage of the available flow. The flow in each canal is then theoretically divided according to the schedule among the various laterals and/or sub-laterals . . . Detailed records are kept showing the available flow and the amount or percentage, going to each of the canals.[1]

The calculation and measurement of water to each diverter along the Chili River is a complicated procedure fraught with many problems. Most of the flow of the Chili River is controlled at the Charcani gaging station with only a small portion coming from the Socabaya River farther downstream (figure 7). In computing the availability of water, the average daily flow as recorded at Charcani is determined. This is done by averaging 3 daily staff gage readings taken at 6 a.m., 12 noon, and 6 p.m. This average daily tabulation affects the amount of water each diverter receives the following day.

Some diverters acquire water on a fixed basis, while others get a percentage of the remaining stream flow. Urban users and one of the principal irrigation projects receive a fixed quantity of water year-round. The city of Arequipa gets 250 l/s supplied through the Zamácola irrigation canal. An additional 250 l/s, allocated to the thermal electric plant located in Chilina, arrives through the Miraflores Canal. In both instances the bulk of this water eventually returns to the Chili River and is used farther downstream beyond the confines of the *Campiña* itself.

Since the privately initiated Zamácola Irrigation Project secured its own water supply from Lake Indio in 1945, it has an assured allotment of 840 l/s. This quantity represents 84 percent of the 1,000 l/s diverted from Lake Indio. By making flow measurements with a current meter at the exit of Lake Indio and again in the canyon of the Chili River, the government estimates that about 16 percent of the original amount captured in the high sierra is

---

[1] Jerald E. Christiansen, "Irrigation and Drainage Problems in Peru, Observations and Recommendations," North Carolina State Agricultural Mission to Peru, Logan, Utah, June 1964, p. 11. (Mimeographed.)

lost to evaporation and infiltration. Two small communities in the vicinity of the Zamácola Project, Cerro Colorado and La Libertad, receive 10 l/s between them and, of course, most of this water returns to the river, to be used by farmers farther downstream.

The remaining river water diverted to the irrigated lands bordering the Chili River is apportioned according to the amount of land under the ditch. Using 1 l/s/ha as the duty of water so widely accepted in Peru was fundamental in this allocation process. These proportions were determined over 30 years ago.[1] Modifications made in the basic allotment for each canal fronting the Chili River in the Campiña took into account the length and physical condition of the canal. The longer canals and those traversing permeable soils with no lining received proportionately the most water. With the spread of urbanization, irrigated cropped land converted to city lots brought about a decrease in water allotments vis-á-vis the standard criterion of 1 l/s/ha.

The manner of allocating water also depends on the time of year. During high stream flow, when there is at least 3,500 l/s in the river, the D.A.R. declares toma libre.[2] Under these conditions diverters use as much water as they are able, but since headgates can admit only a limited quantity of water, diverters downstream are not put at a disadvantage in receiving a fair share of this resource. Paradoxically, during this season, it is not uncommon for farmers to be confronted with a water shortage brought about by the washout of their small diversion dam in the river. This leaves the headgate high and dry until replacement of the dam occurs.

During the dry season when there is less than 3,500 l/s in the river, diverters in the Campiña receive a percentage of the flow in the Chili River (table 27).[3] Within the Campiña, the farthest upstream canal receiving water in this manner is Acequia Alta, and the last downstream canal is Chichas, where the farmers have the right to diver all the remaining water in the river, an act referred to as tapar al seco.

Diverters farther downstream from the Chichas Canal depend on return flow. The first major diverter after Chichas is Tiabaya, which, during the dry season, can divert all the stream consisting

---

[1]La Dirección General de Aguas de Regadío del Ministerio de Agricultura, Arequipa, interviews with employees, 1966.

[2]High stream flow usually occurs from December through April, see table 22.

[3]Benjamín Caro Saavedra, "Reglamentos de Distribución de las Aguas de Regadío en los Valles de la Costa," Informe de la Inspección General Técnica de la Dirección General de Aguas de Regadío del Ministerio de Agricultura, Lima, Perú, 1966. See table "Valle de Arequipa." (Mimeographed.)

TABLE 27

DISTRIBUTION OF WATER FROM THE CHILI RIVER DURING THE DRY SEASON, CAMPIÑA OF AREQUIPA, JULY 25, 1966

Date of Stream Flow Measurement at Charcani Station	Time of Measurement			
	6 a.m.	12 noon	6 p.m.	Average for day
24 July	3,961 l/s	3,671 l/s	5,387 l/s	4,340 l/s
25 July	4,187 l/s	4,276 l/s	5,233 l/s	4,562 l/s

Canal	Water Right (%)	Water Right (l/s)	Discounted (Urbanization) (l/s)	Final Allotment (l/s)
A	B	C	D	E
Chilina	0.7	25	0	25
Acequia Alta	25.0	875	30	845
Miraflores	8.0	280	74	206
Mantilla	4.5	157	10	147
San Francisco	1.5	53	10	43
San Juan de Dios	4.8	168	30	138
San Jeronimo	6.0	210	20	190
Chullo	6.0	210	10	200
Antiquillo	14.0	490	30	460
El Medio	5.4	189	10	179
La Polvora	2.0	70	15	55
Tio	11.0	385	0	385
Sachaca	6.6	231	11	220
Chichas	4.5	157	0	157
Potable Water	----	---	--	250
Total	100.0	3,500	250	3,500

SOURCE: Dirección de Aguas de Regadío, Arequipa.

NOTE: Total flow to be distributed in the Campiña of Arequipa for July 25 is 4,340 l/s which was obtained from averaging the previous day's flow. With a fixed supply for the Zamácola Irrigation project of 840 l/s, this leaves 3,500 l/s to be distributed to the remaining 10 canals. The canals of Miraflores, Mantilla, San Francisco, San Juan de Dios, and San Jeronimo have one headgate named Miraflores. Column D represents the amount of flow subtracted from the original water right indicated in Column B as a result of partial urbanization of farm land.

of return flow. Between this canal and Socasani where the headgate for the La Joya Irrigation Project is located, there are several small irrigated areas along the Chili River that are largely uncontrolled. The average amount of water allotted to the La Joya Antigua Irrigation Project is 3,389 l/s, although near the end of the low stream flow period, it averages only 2,179 l/s.

In order to control the amount of water each canal receives during the dry season, the D.A.R. assigns a controller of the river, controlador del rio, to regulate the headgates of all diverters upstream of Tiabaya. Each of the ditches also has a controller of the headgate tomero, whose duty it is to keep in force the setting ordered by the D.A.R. The tomero is under the direct control of the individual users he serves. The controlador del rio, while financially supported by all diverters, is under the supervision of the D.A.R. Based upon the average stream flow as recorded at Charcani for the previous day, a list is prepared indicating the amount of water each ditch is to receive. The controlador del rio sets out early in the day, starting at the uppermost headgate and ending at Chichas, adjusting the flow into each headgate according to the list. In the afternoon he travels upstream rechecking the headgate settings. Only a few of the ditches contain a Parshall flume which permits reasonably accurate flow measurements. Most ditches use a portable rectangular weir which leaves much to be desired in arriving at accurate allocations.[1] When Parshall flumes are used, it is much easier to guard against tampering with the settings.

Two additional problems in allocating water became obvious from observations in the area. The first concerns the basis of allocating water during the dry season among the ten canals getting a percentage of the river flow. The allocations are not based on the present day's stream flow but on the previous day's flow, with no night measurements being taken (table 27). Calculating the water losses in route to the Zamácola Irrigation Project presents another complication. Even with the flow measurements made, accounting for losses due to infiltration and evaporation, only gross estimates result if for no other reason than that little consideration was given their variations through time.

With the functioning of the reservoirs in the high sierra,

---

[1] Roland C. Holmes and Benjamín Lizaraso C., "Informe sobre Reconocimiento del Estado Actual de Captación de las Aguas del Rio Chili en la Campiña de Arequipa," A report to the Director del Programa la Joya, Ing. Otto Schultz Wetzell, Arequipa, Perú, 1966. (Typewritten.)

which did not occur until this study was nearing completion, the manner of allotting water to diverters changed so that toma libre does not exist and each diverter receives a fixed daily allotment. In the Campiña, the ten canals that formerly received a percentage of stream flow during the dry season now receive a fixed percentage year-round. Now the calculations are based on 4,000 l/s instead of 3,500 l/s. In addition, the canal of Tiabaya has a limit on its intake of river water under the new allotment plan. To accomplish this, the D.A.R. had a new headgate incorporating a Parshall flume built to replace the rustic headgate. Before this was used, it took time to convince farmers of the fairness of the new plan and that an assured flow of water could be guaranteed year-round. In fact, the new headgate lay idle for several months before the farmers cooperated with the government and built a diversion dam to funnel water into the headgate. Also, according to the new plan, those remaining headgates in the Campiña with portable rectangular weirs will be replaced with Parshall flumes. In order to supervise this new allocation plan, and instead of having temporary tomeros for each canal, there now is a permanent tomero for each, still to be paid by the junta de regantes of the respective canal. Replacing the controlador del rio is a chief in charge of allotting water in the Chili Basin (jefe de repartir), who checks on the settings of each headgate. This might occur only once every two or three days. The amount allocated to the various canals in the original La Joya Irrigation Project is carefully controlled so as to assure San Isidro an allotment of 1,500 l/s during the initial stage of colonization.

Apart from developing a new allocation system for the Chili Basin, the government had to contend with many demands for an increase in allotment, once water began flowing from the new reservoirs. The situation was especially acute in the original La Joya Irrigation Project, where only about four-fifths of a farm was irrigated, partly due to a lack of water. These farmers petitioned the government to have their water quota increased to 1 l/s/ha year-round, so that they could cultivate all of their farms. They believed that, before any new project such as the extension of the La Joya Irrigation Project was undertaken, older projects should first be fully developed.

Additional pressures were placed on the government during the initial days of colonizing San Isidro. At this time when the delivery of water was so sporatic and uncertain, the new farmers of the project complained bitterly that others, especially those in the original La Joya Irrigation Project and even farmers in the Campiña, were taking more than their rightful share of water. On

TABLE 28

DISTRIBUTION OF WATER FROM THE CHILI RIVER AS RECORDED BY DIFFERENT AGENCIES, JULY 25, 1966[a]

Canal	Measurements (l/s)	
	DAR	ONRA
A	B	C
Tio	385	745
Sachaca	220	237
Polvora	55	108
Acequia Alta	845	1,279
Miraflores Complex	724	831

SOURCE: Field Survey and records of the Dirección de Aguas de Regadío, Arequipa.

[a]Only ditches with Parshall flumes were examined here. Compare data in column B with Column E, table 27, p. 88. Only water for agriculture use was considered. In every case by an independent examination of the relevant Parshall flumes by ONRA, these ditches were receiving more water than recorded in the official files.

several occasions a delegation of these new farmers made unexpected trips to the original La Joya Project and to the Campiña, hoping to observe irregularities and report them to the authorities. Table 28 gives an example of improper allocations of water occurring in the Campiña during this transitional period. While new sources of water open up possibilities for the expansion of irrigated areas, opportunities for rectifying existing deficiencies in management practices also occur, taxing the ingenuity of the government to meet the demands of farmers who now feel entitled to a fair share of water. International lending agencies are aware of these problems. In considering a loan for the Aguada Blanca Dam Project in the upper Quilca Basin, the Interamerican Development Bank (IDB) recommended the following in a working document:

1. that Peru install appropriate water measuring instruments such as Parshall flumes at the headgate of each ditch along the Chili River so as to systematically control the water below Aguada Blanca,
2. that there be one agency in charge of managing water distribution and operating the dam system in the future, and
3. that an irrigation ordinance be developed resulting in a "correct" operation

of the hydraulic system in the Chili Basin. The ordinance should contain information as to:

a. an annual time table of cropping and irrigation,

b. duties of farmers (irrigators),

c. duties of the governmental institution that operates the hydraulic system

d. personnel needed to distribute water,

e. sanctions to use for infractions,

f. annual budget, and

g. other items.

Thus, the results of augmenting the water supply indicate that society must learn to balance the distribution of water, not only for present users, but also for succeeding generations.

CHAPTER IV

IRRIGATION WATER CONTROL IN AREQUIPA[1]

An essential part of all irrigation schemes entails moving water from its source to the farm and on to the fields. This operation reflects the interplay of the physical environment, the perceptions of this environment by farmers and the technical-scientific community alike, and the use made of perceived opportunities as these three variables affect the choices arrived at in fashioning an irrigation system.

In this chapter the management of water at the farm level is discussed after outlining the nature of the water conveyance system. The description of this system portrays its physical design, operation, and maintenance. This delineation centers on the Chichas Canal located in the *Campiña* with some comparisons to the projects of La Joya Antigua and San Isidro on the Pampa of La Joya (figure 2). The following assessment of on-farm mangement of water covers the farmer's knowledge and perception of the water supply, the methods used in watering the field, the nature of scheduling irrigation waterings, the amount of water applied, and the manner of transferring water resource knowledge to the new irrigation project of San Isidro. All quantitative data describing conditions in the study area come from actual field measurements.

As indicated in chapter II, farmers receive insignificant amounts of technical aid in the management of irriagion water. Two notable exceptions, though, occur: only under the supervision of the *Dirección de Aguas de Regadío* (DAR) can modifications of headgates and community ditches take place; and during the early colonization of San Isidro, the development of the complete irrigation system, including on-farm management of water, was under the control of the

---

[1]The title of this chapter comes from an idea expressed by C.D. Criddle who states that

> "With any type of irrigation control is the keyword. Control of water is possible with equipment and only limited labor or it can often be obtained using adequate labor and only limited equipment . . ."

This quote appears in Jack Keller, "Effect of Irrigation Method on Water Conservation", *Journal of the Irrigation and Drainage Division, Proceedings of the American Society of Civil Engineers* 91 (June 1965): 71.

Oficina Nacional de Reforma Agraria (ONRA). Therefore, as a surrogate for the general absence of a stance taken by the local engineers on optimal solutions to water management problems in the study area, reference is made to general technical literature. The degree to which local engineers' views correspond to these guidelines will be indicated where appropriate.

The Conveyance Network

Getting water from the source to the farm is fraught with many problems. Chief among these is the manner the conveyance network fits into the landscape as denoted by the length, density, symmetry, and slope of the canals. Additional important aspects of the network's physical design entails the type of lining employed and the manner of structurally regulating the flow of water within a single community ditch system through the use of farm ponds. Also, the way water delivery occurs and the canal system is maintained present problems and opportunities for the resource manager.

The Physical Design of the Conveyance Network

In figure 8, one can readily appreciate the pattern of the water distribution network in the zone of Chichas. Zimmerman has noted that a well-designed irrigation ditch network should appear to be simple and a part of the natural landscape, giving the impression that the constructed network is the only logical solution under the existing local conditions.[1] But as a consequence of various factors - the lay of the land, the size and shape of farms and fields, and the land tenure system - there has resulted a seemingly fortuitous ditch network pattern in the Chichas area consummated in innumerable small ditches that distribute water from the main canal to farms and fields.

The total length of all ditching in the Chichas area is almost 30 km, consisting of 5,650 m of main canal, 12,730 m of laterals, and 10,770 m of farm canals.[2] Furthermore, the ditch network covers approximately 54,000 m^2, with 13,560 m^2 in main canal, 22,914 m^2 in laterals, and 17,232 m^2 in farm canals. These calculations assume an approximate width for the main canal to be 2.4 m (1.4 m for the

---

[1]Josef D. Zimmerman, Irrigation (New York: John Wiley and Sons, Inc., 1966), p. 203.

[2]Though both the main canal and laterals are community ditches, the latter commonly supplies water directly to individual fields. Thus, for certain calculations, data for laterals and private farm ditches are aggregated.

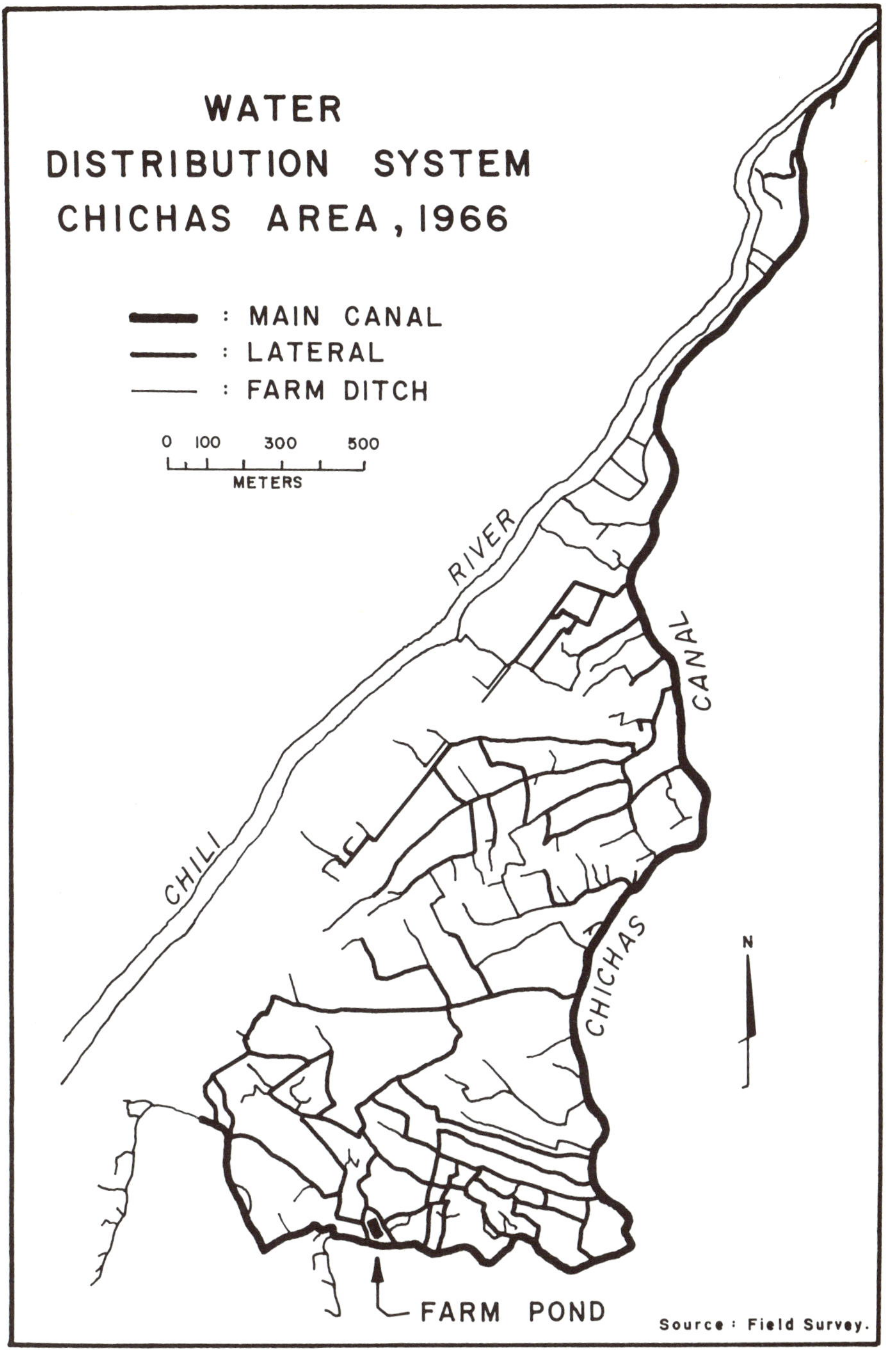

Fig. 8 - Water Distribution System, Chichas Area, 1966.

bed of the canal and 1 m for the 2 borders), 1.8 m for the laterals (1.0 m for the bed of the canal and 0.8 m for the 2 borders), and 1.6 m for the farm canals (0.8 m for the bed of the canal and 0.8 m for the 2 borders). This arrangement results in almost 4 percent of the 135 ha under the command of the Chichas Canal in ditches; whereas, combining laterals and farm ditches, the area occupied by ditches is 3 percent. The overall density of the distribution network is 216 m/ha of land cultivated while considering jointly the laterals and farm ditches it is 174 m/ha and only 42 m/ha when considering only the main canal.

Askochensky also made observations about the dense conveyance network in the arid zone of the U.S.S.R. before the Revolution.[1] He suggested improving this situation by consolidating the many small ditches into a few major canals which would result in less area occupied by ditches and more area in cultivation. With this modification less water is lost through infiltration bringing about a higher conveyance efficiency. As Zimmerman observed, if spacing between ditches is doubled, it naturally results in a diminution by halving the number of ditches.[2] The worst that could occur from doubling the ditch capacity would be a modest increase in the cost. With the doubling of the number of ditches though, there almost always results a doubling of the cost of the ditch network. A realignment based upon Zimmerman's argument appears to be a solution to the improvement of the Chichas conveyance system, but since this area, as well as much of the Campiña, is highly broken and terraced (figure 9), the consolidation of the smaller ditches into larger ones would be physically difficult.

In the Campiña the conveyance system was laid out in an uncoordinated manner. Each succeeding area brought under the ditch was fitted into the landscape generally to abut the higher adjacent irrigated land. However, this procedure was not followed in the projects on the Pampa of La Joya. The conveyance system of projects, planned before being built, was developed in an integrated fashion. Also due to less rugged terrain here, both farms and fields form a rectangular or square pattern. Thus, on the Pampa of La Joya a more orderly array of ditching exists than in the Campiña.

The ditch information pertaining to La Joya Antigua refers to

---

[1]A.N. Askochensky, "Basic Trends and Methods of Water Control in the Arid Zones of the Soviet Union," in The Problems of the Arid Zone (Paris: UNESCO, Arid Zone Research No. XVIII, 1962), pp. 401-402.

[2]Zimmerman, Irrigation, p. 216.

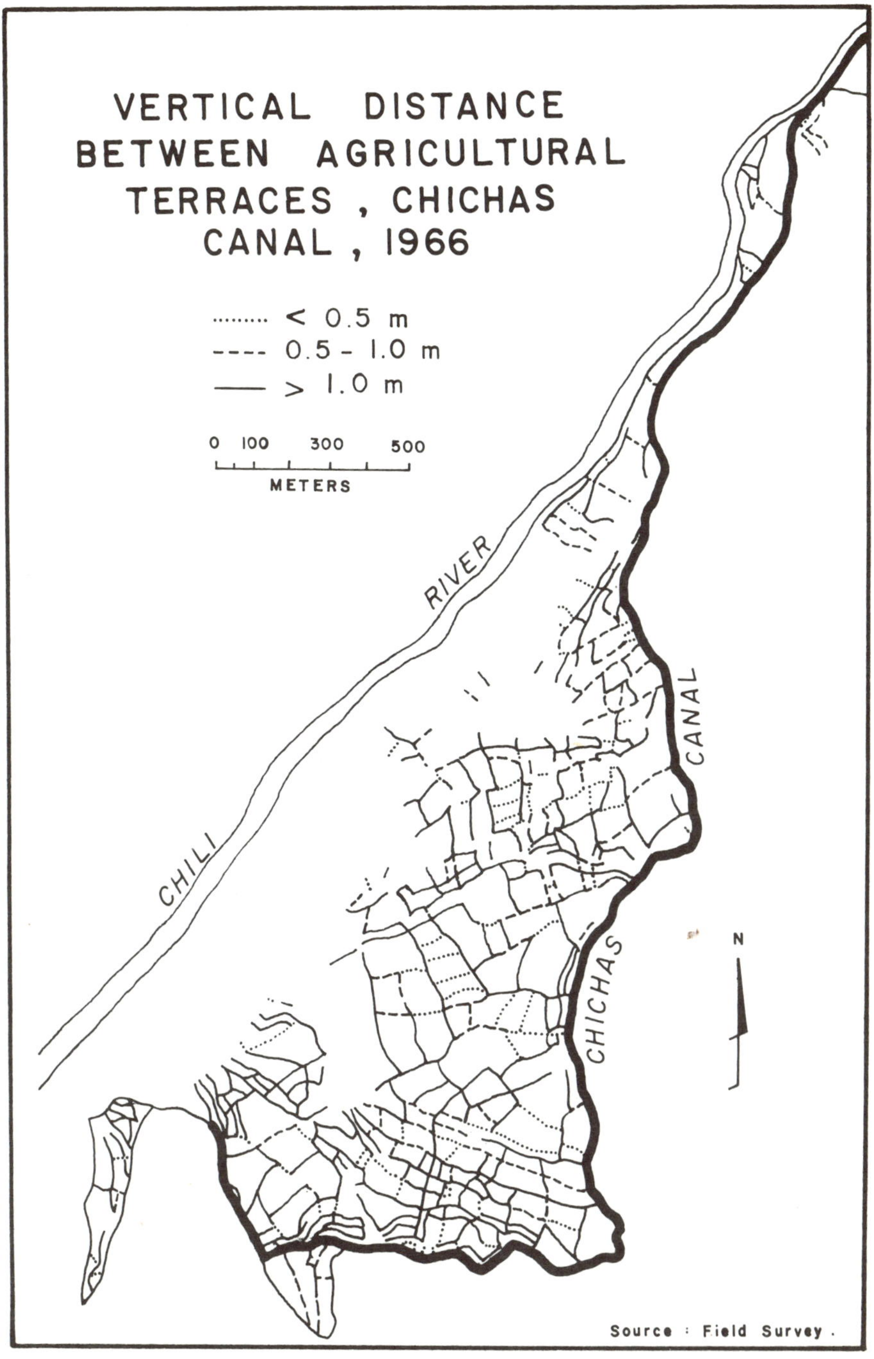

Fig. 9 - Vertical Distance between Agricultural Terraces , Chichas Canal , 1966 .

the distribution network as found on 14 farms, having an average size of 5 hectares. Ranging from 1.1 to 3.8 percent, approximately 2.2 percent of these farms is occupied by ditching. The basis for these calculations assumes the average width of the farm canal to be 1.7 m (0.9 m for the bed of the canal and 0.8 m for the 2 borders). The density of the distribution network within these farms is 130 m/ha.

In San Isidro, ditch information pertains to the canal network of the area under cultivation along Lateral--B, the first and second stages of the physical development of 55 farms each having a developed area of 7 hectares. Again, only on-farm ditching is under examination. The percentage of the area occupied by the ditch network within the 55 farms is 2.3 percent, slightly greater than for La Joya Antigua. This difference may stem in part from assumed wider canals of 1.8 m (0.8 m for the bed of the canal and 1 m for the 2 borders) in San Isidro than in La Joya Antigua. The density of the ditch network is 128 m/ha.

Considering the above data and general field observations, the irrigation projects on the Pampa of La Joya have a slightly smaller amount of land in ditching than in the Chichas area. The density of the canal network in San Isidro is less than that in La Joya Antigua and certainly than that for the zone irrigated by the Chichas Canal.

The symmetry of the area commanded by a ditch reflects in part the meshing of ideas which engineers and farmers hold. As Zimmerman indicates, an ideal traverse for a canal that allows it to irrigate on both sides is along a ridge.[1] Nevertheless, the main Chichas Canal, throughout most of its course, commands only the right side (figure 8) since the Polvora and other canals that command the area to the left of Chichas were presumably constructed first. Likewise, the area commanded by the principal canals in the Campiña is generally along one side of the ditch.

A common criterion used for laying out a canal is to follow the contour of the land. This makes possible the irrigation of the maximum surface. But for land that is hilly, following contour lines results in an excessively long canal with many curves. In order to avoid this situation, it is necessary to employ tunnels, filled sections, flumes, inverted siphons and stable chutes or drops to shorten the length of the canal. Several of these structures have been incorporated along the Chichas Canal. One small

---

[1] Ibid., pp. 217-218.

flume carries this canal across a small stream while another flume carries it over a railroad track (figures 10, 11, and 12). Since the depth of the small stream and the railroad excavation is about 10 m each, flumes are probably the most economical structures. But, when a stream valley has a depth of 30 m or more, the most economical means of traversing it is by a reinforced-concrete inverted siphon.[1] There is an example of this in the Majes River Basin near Ongoro. It is also convenient to use this type of structure in crossing a wide road. An inverted siphon carries the main canal of Chichas under the paved highway linking the city of Arequipa with the town of Tingo (figure 13).

Another critical factor affecting the proper functioning of a canal is its gradient. Selecting the optimum grade is difficult since, to command the most land by a canal, the slope should be as gentle as possible but not less than 0.5 percent. A slope with less than this amount results in costly and tedious construction. Furthermore, with time, a canal with an extremely gentle slope facilitates the growth of aquatic plants and the accumulation of sediments.[2] However, if the canal has a steep slope, the result is a swift and uncontrollable current that causes erosion of the walls and bed of earthen lined canals. And from an economic point of view, a canal designed for high velocities is most costly to construct in spite of the fact that its cross-section is relatively smaller in relationship to canals with more gentle slopes. The saving that results in the smaller cross-section is counterposed by the structures used to decrease the pressures at each headgate.[3]

The longitudinal profile of the principal canal of Chichas has a slope that varies greatly from extremely gentle (0.22 percent) to very steep (4.06 percent) (figure 14). In a portion of section, G-H, with an average slope of 0.22 percent, the gradient of the bed of the canal actually increases rather than decreases as the water flows across country and causes ponding (figure 15). On the other hand, the section, H-I, has a slope of approximately 4.06 percent and, in order to prevent erosion, farmers lined the sides and bed of the section with large rocks and boulders (figure 16). Only in this latter situation was corrective action taken.

In many instances, for the projects on the Pampa of La Joya abrupt or steep gradients of canals do not present grave situations

---

[1] Edward Kuiper, Water Resources Development (London: Butterworths, 1965), p. 360.

[2] Zimmerman, Irrigation, p. 216. [3] Ibid., p. 221.

Fig. 10. A reinforced concrete flume of the Chichas Canal crossing a small intermittent stream (<u>torrentera</u> or <u>lloclla</u>). The width of the canal narrows as it approaches the flume.

Fig. 11. Side view of flume (See Figure 10). A concrete post supports the middle of this structure.

Fig. 12. A leaky wooden flume crossing a railroad cut in the Chichas area. The water flowing beyond this flume irrigates only one farm.

Fig. 13. Inverted siphon along the Chichas Canal. The canal crosses under a main road and railroad line at this point near the city of Arequipa.

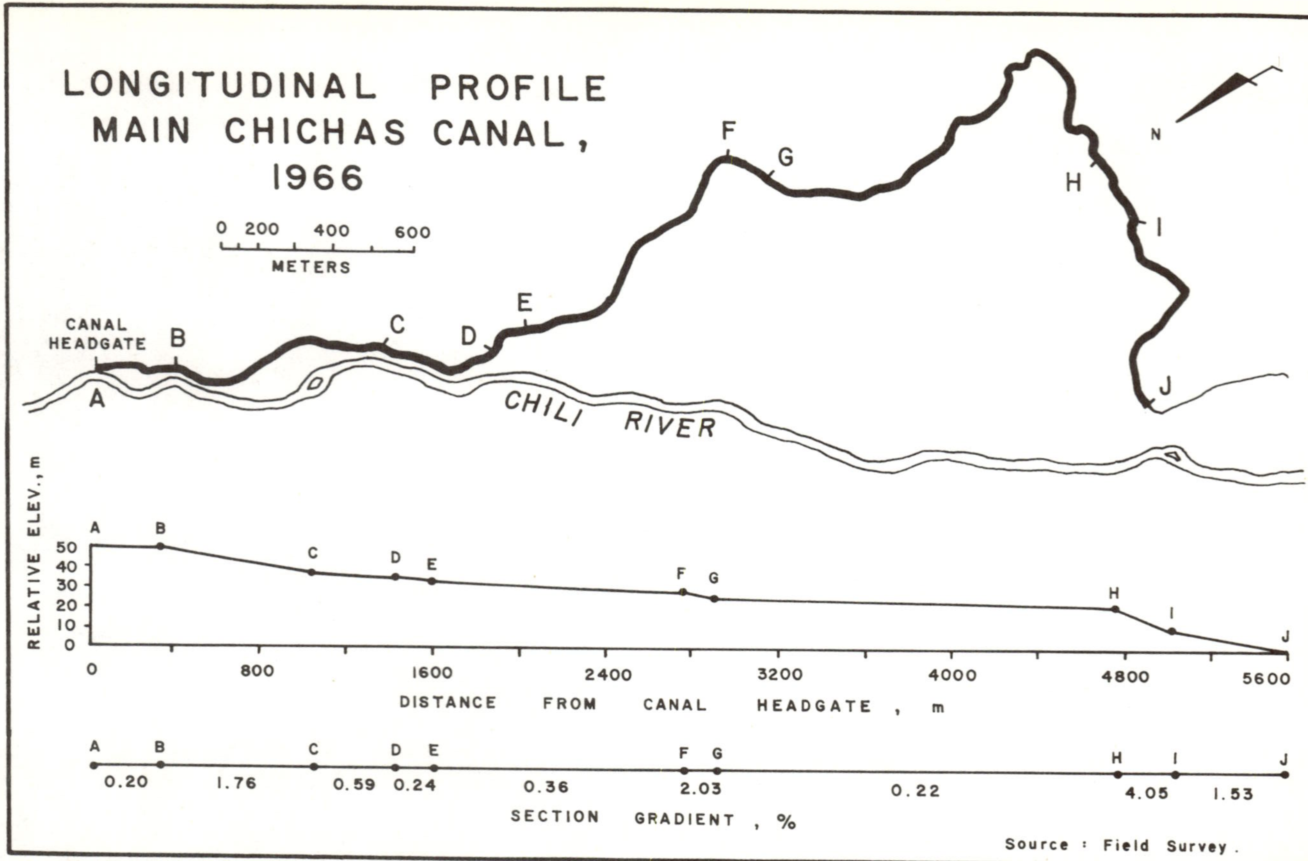

Fig. 14 - Longitudinal Profile of Main Chichas Canal, 1966.

Fig. 15. A gentle gradient along Chichas Canal. In this section of the canal the water overflows the canal banks and covers the exposed land in the foreground.

Fig. 16. A section of the Chichas Canal with a pronounced slope of over four percent. Here the farmers have lined the bed with rocks to avoid serious erosion.

that have not already been resolved by the engineers during the initial construction period. For example, in La Joya Antigua engineers have placed chutes to drop water from laterals to lower elevations with stilling basins designed into the system. However, in San Isidro high uncontrollable velocities occur along the primary laterals due to the interplay of appreciable longitudinal slopes and a telescopic form of canal construction. As a result, difficulties arose in maneuvering the portable wooden checks used to divert water to sub-laterals. Possibly, as voiced by some of the local engineers, if the grooves into which the checks were inserted had been slightly inclined downstream rather than vertical, the checks might have been easier to maneuver. Then an additional problem demanding quick action was correcting for overflow at the sub-lateral headgates. ONRA had to raise the height of the canal walls in the immediate vicinity of these junction points to control the overflow (figure 17). These are two of several examples cited in this chapter when designers and builders (usually civil engineers) of the irrigation conveyance system have not been cognizant of the agricultural needs in general or of the water-use practices in particular of farmers and agricultural engineers.[1]

Canals that are not lined present many problems, yet farmers in the Chili Basin believe that unlined ditches are economical when in reality they can be extremely costly. Zimmerman enumerated unsuspected costs allied with unlined canals.[2] Although the main canal, laterals and sub-laterals are concrete-lined in the projects on the Pampa of La Joya, only a small section, about 100 m of the main canal of Chichas, is lined and covered. This recent portion of the canal is near the new neighborhood of Los Alamos (figure 18).

Since most of the Chichas Canal is not lined, there are two pressing problems apart from the loss of much water by infiltration (figure 19). Near the head of this canal, where it passes several rural residences, cave-ins that have caused significant amounts of residential damage have occurred. A concrete wall now protects the zone (figure 20). Also, along other sections of this canal there have been additional cave-ins with the government repairing the damage and those farmers affected covering the cost through a special levy. Such cave-ins also adversely affect the scheduling of irrigation water, and when the time taken for repairs is lengthy,

---

[1]United Nations, Food and Agriculture Organization and the International Bank for Reconstruction and Development, The Agricultural Development of Peru, part II: Detailed Report, Water Utilization, pp. 132-133.

[2]Zimmerman, Irrigation, p. 278.

Fig. 17. A precarious approach to a turnout along Lateral - B in San Isidro. In order to avoid loss of water and erosion of canal banks, engineers raised the walls in a makeshift manner using rocks.

Fig. 18. Cement lined canal. Along the left bank of the Chili River, the Chichas Canal is cement lined and a sidewalk has been constructed over the canal adjacent to the new housing project (<u>barrio</u>), Los Alamos. Much of the remaining portion of the canal remains unlined.

Fig. 19. Evidence of canal seepage. Loss of water occurs through infiltration from the unlined Chichas Canal located in the upper left corner.

Fig. 20. A needed reinforced canal wall. A cement lined wall of the Chichas Canal protects the adjacent rural homes from inundations.

it can have devastating effects on crops (figures 21 and 22).

A second problem with unlined canals involves the vegetation growing along their borders as seen adjoining the Chichas Canal (figure 23). Apart from impeding the flow of water, the vegetation itself consumes large quantities of water. Yet most farmers do not consider the water lost since bank vegetation, consisting of grasses and trees, help to stabilize the sides of canals; also, these solitary trees add aesthetically to the landscape and provide shade for the passerby.

A useful structure comprising a portion of the conveyance system is the farm pond. These small reservoirs exist throughout the Campiña, one being near the end of the Chichas Canal (figures 24 and 25). This tank has an approximate capacity of 7,500 m^3 commanding 15 ha of a 30 ha farm. Booher mentions several advantages in the use of these small reservoirs as permitting daytime irrigation, reducing irrigation hours, and increasing the efficiency in the use of water.[1] By combining upstream tailwater collected in the farm pond with the regular water-right allotment, a larger stream allows the owner of this reservoir to irrigate more effectively, including the avoidance of nighttime irrigation.

Water Delivery and Its Supervision

As Hudson aptly expresses it, though in a somewhat different context, "The only real goal in delivering water is not that it should be distributed efficiently or fairly, but that the farmers should believe that it is being distributed fairly, thereby reducing the amount of 'murder or suicide' in the hearts of members . . ."[2] of an irrigation community. By and large the delivery system for the Chili Basin fulfills this goal. It is only during critical times, especially when the stream flow is very low, that many complaints are received by the DAR in Arequipa necessitating visits by government personnel to the field in order to resolve problems of water distribution.

Throughout the entire area, the dominant mode of delivering water is by rotation. In general, the frequency of delivery is less in the Campiña than on the Pampa of La Joya. Farmers in the Chichas area receive their turn (mita) at intervals between 15 to 16 days,

---

[1] L.J. Booher, Surface Irrigation (Rome: Food and Agriculture Organization of the United Nations, 1967), p. 6.

[2] Hudson, Irrigation Water Use in the Utah Valley, Utah, p. 136.

Fig. 21. A fragile canal wall. Government personnel and concerned farmers inspect a weak bank of the main Chullo Canal in the *Campiña* of Arequipa.

Fig. 22. Repairing a main canal wall cave-in. Government personnel supervise this work taking place in the Campiña of Arequipa.

Fig. 23. Much vegetation grows along the main Chichas Canal, which remains mostly unlined.

Fig. 24. Farm pond located near the end of the Chichas Canal. Trees and brush ring this reservoir.

Fig. 25. The outflow from a farm pond (See figure 24). Large blocks of rock constitute this well fortified part of the reservoir.

in the Zamácola Irrigation Project every 7 days, and in La Joya Antigua every 3 days. This rotation plan does present hardships for the farmer. He receives on a rotation basis one watering during the day and the next one at night. In the Zamácola Project it even forces the farmer to receive his mita always on the same day of the week.

Generally the duration of a turn is related to the size of the farm; the larger the farm, the longer the turn. However, the length of time of a turn normally does not permit a farmer to irrigate his entire farm. Therefore, a farmer has to decide which fields and which crops will be irrigated at each turn.

During the initial stage of colonization of San Isidro, no fixed rotation cycle was arranged other than to have one delivery arrive during the day and the succeeding one at night. Since farmers were guaranteed 100 l/s, the amount of water flowing into the project determined the time between irrigations. Sometimes this lengthened to more than a week, causing droughty conditions, especially for crops not well rooted.

Normally the supervision of the delivery of water to farmers is handled by farmers themselves. In the Campiña and in particular in the Chichas area and the Zamácola Irrigation Project, the farmer will usually inspect farm headgates upstream from him before beginning to irrigate. If the flow is exceptionally low, as it might be during low stream-flow periods, a farmer during his turn may make additional inspections upstream. Likewise, farmers during their night turns are especially cognizant of the need to keep a continual watch of upstream conditions along the conveyance network.

This surveillance system is different on the irrigation projects on the Pampa of La Joya. In La Joya Antigua and in San Isidro the government ditch riders aid the farmers by supervising the opening and closing of farm headgates.

Physical Maintenance

Since many canals in the Chili Basin are old and in poor physical condition, constant maintenance is needed. In maintaining and improving the conveyance system, engineers are not alone in accomplishing these tasks.

Usually any modification to community ditches or structures adjoining them must not only meet the approval of the appropriate governmental agency, but, just as importantly, the approval of the farmers affected. To illustrate the importance of farmer approval, a government engineer, thinking a revamped old automatic divider would be a technical improvement, built a new one at the request of a few farmers only to find out that a grievous error had been committed. The results were disastrous. Since this automatic partitioner affected two main irrigated areas, Pampas Nuevas and Tiabaya, farmers of the latter area, believing they had not been properly consulted before the construction, were not surprised to learn that the new structure had been destroyed the night after it was opened for use. This episode brought the National Director of the DAR from Lima, a police escort, and many affected farmers to the scene of trouble. Finally it was decided that a divider similar to the original one would be reconstructed under governmental supervision.

Another example of a delicate situation where the flow in a community canal was affected occurred along the Tiabaya Canal. Here a farm turnout collapsed along a weak and steep bank. In order to replace this structure, the affected downstream farmers and the government had to reach an agreement on reestablishing the original flow along the damaged portion of the canal (figure 26).

Innumerable examples seen by the author portray the delicate automatic partioning of flows along community ditches in many areas, not only in the Chili Basin but also in the ancient irrigated areas of Chivay, Chuquibamba, and elsewhere in South Peru. Specifically, use is made in many of these areas of large boulders placed decades ago in the community ditch. These serve as an automatic partitioner of water to two or more areas simultaneously. Normally these are not cemented in place and their positions can change from time to time. However, in one instance, engineers working along the Paucarpata Canal in the Chili Basin wanted to replace the large stones with a permanent, well-designed divider, but the affected community of farmers objected for fear that someone's water rights would be adversely affected.

Water Management on Farms

As already mentioned in the first chapter, one of the most neglected phases of irrigated agriculture is on-farm water management. The questions discussed in this section are of fundamental importance in understanding not only the present functioning of irrigation projects but also the problems and opportunities related to the improvement of existing projects and the development of new ones. These questions are:

1. What is the farmer's knowledge of the water supply?
2. What are the sources of water?
3. How do farmers apply water to their fields?

Fig. 26. Initial stage in repairing a canal cave-in. Measuring the flow of water with a Price current meter along the Tiabaya Canal just upstream from the damaged farm turnout along the left bank. Government officials and concerned farmers observe this operation.

4. When do farmers apply water?
5. How much water is used in the watering process?
6. How is water resource knowledge transferred?

Farmer's Knowledge of the Water Supply

A prerequisite for a farmer in applying the appropriate amount of water is knowledge about the quantity that he receives. From observations and discussions with farmers, few can state the quantity they receive; they can only indicate the duration of their turn. Since great fluctuations occur in the available supply, this information is of little use to a farmer trying to anticipate his irrigation schedule which is based upon the amount he receives at a future turn. Furthermore, without the aid of irrigation extension agents, who are scarce in the Chili Basin, farmers cannot make practical use of volumetric or rate of flow data even if available. In general, farmers appear to be uninterested in this type of information. In fact, on those farms where the author measured the amount of water applied at each watering, farmers never asked about the amount of water used by them. Generally, these findings from Southern Peru substantiate those found by Hudson in the Utah Valley where most farmers could only state the quantity of water received in units of time.[1]

One exception to this generalization occurs in San Isidro, where engineers designed the irrigation system so that farmers would automatically receive 100 l/s as noted previously. Preceding this arrangement, ONRA had to remove the original small measuring flumes incorporated in the farm headgates which allowed only a maximum flow of about 50 l/s.

In another instance, a farmer seriously considered measuring the rate of flow into his farm and also the rate of soil moisture used by his crops. In the Santa Rita de Siguas Irrigation Project outside the study area but adjacent to the Pampa of La Joya, an engineer, faced with an acute shortage of water on his farm, obtained through the writer a set of tensiometers to measure changes in soil moisture. After a few weeks of use, this equipment was abandoned. The overseer (mayordomo) of the farm was not able to use it properly. This undertaking was remarkable considering no formal aid emanated from the government. Surely this must have been one of the first times a farmer in South Peru had tried to put into operation a somewhat sophisticated system to economize on the use of

---

[1]Ibid., p. 147.

a limited water supply. To be sure, the large sugar cane haciendas in the neighboring Tambo Valley have, for several years, been using advanced techniques. But these farms are large operations with technicians especially trained in soil-plant-water relationships.

An important aid for farmers trying to determine a yearly planting program is the availability of information on future water supplies. In the Chili Basin this type of data is practically nonexistent. Nevertheless, some local technicians offer information about future water supplies to farmers. Perceiving the existence of a climatic pattern for the local area that results in a 7-year stream flow cycle, technicians attempt to forecast the coming of wet and dry years. But nothing published for the local area proves their point.

Some farmers use another method to gauge future water supplies. By observing the variations in the snow line on the adjacent mountains of Charcani, El Misti, and Pichu Pichu, they believe this empirical evidence helps them plan for the future. With the occurrence of a good snow pack on El Misti late in the season, the author has heard several farmers voice with optimism the coming abundance of stream flow during the normally low stream flow period. Yet, as indicated in the previous chapter, melt water does not greatly influence the average monthly stream flow.

Although farmers may be imprecise in stating the amount of future flow of water received, many have a definite notion about the quality of water for irrigation. Farmers prefer river water over spring water. They are, however, cautious in their use of river water heavily charged with silt when a newly planted alfalfa field is to be irrigated, inasmuch as a layer of silt covering the field makes it difficult, if not impossible, for germinating alfalfa plants to break through the crusted surface.

In many areas in the Chili Basin homes bordering irrigation ditches channel their sewage into them. The Chichas Ditch is no exception (figure 27). Contrary to the views of health officials, farmers perceive this type of polluted water as more valuable for irrigation than river or spring water. Even though this contaminated water carries organic material useful to plants, authorities hope to put an end to its use for irrigation by requiring all residential sewage to be treated at the waste water treatment facility in the city of Arequipa.

A water quality problem that generally does not exist in the Chili Basin is salinity. River water in the area is low in salts. Yet, in one of the study areas, it was high enough to affect the manner farmers use water. With the passage of time in La Joya

Fig. 27. Many canals are used in discarding sewage. The bottom part of a sewer pipe from a home along the main Chichas Canal is visible in the center.

Antigua, farmers have expanded the cultivated area through the use of irrigation runoff. Since this runoff is saline, farmers, in order to grow crops, alternate waterings between regularly allotted river water of low salt content and saline-charged runoff water held in a farm pond. Under these circumstances, farmers raise moderately salt-tolerant crops such as onions and alfalfa.

Source of Water

Almost all farmers in the Chili Basin rely on surface water, both canal and runoff, as their only source for irrigation. In a limited way, some contiguous fields receive water from both sources. For example, one Chichas farmer has a small make-shift flume that funnels runoff from a portion of the Polvora area directly to his farm. It appears this farmer has inferior water rights to the Chichas Canal.

Naturally, if a farm is greatly fragmented, the sources multiply. About a fifth of the farmers in the Chichas area cultivate land in other places as far away as Ongora in the Majes Valley. Farmers in the *Campiña* who also farm land in La Joya Antigua or elsewhere, usually have a permanent worker (*mayordomo*) on these farms. With a *mayordomo* in command, the task of irrigating a fragmented holding is, to a degree, under control.

Though a *mayordomo* may be an experienced irrigator (*camayo*), farm fragmentation presents problems in coordinating farm operations. A contiguous grouping of fields is the ideal farm pattern for allowing better control of waterings and other farm operations. Authorities recognize the merits of the consolidated farm, and Peruvian agrarian reform programs, including the present one, attempt to solve farm fragmentation, but only with some success.

Irrigation Methods

Using the nomenclature of Hudson, three facets of irrigation are understood.[1] These are: diversion of water to fields, preparation of the land, and application of water.

Diversion of water to fields. In the *Campiña* and the irrigation projects on the Pampa of La Joya most farmers transport water to fields by gravity flow. When the idea of using small pumps and portable piping to irrigate land not commanded by a ditch was

---

[1] Ibid., p. 162.

suggested, generally the farmers rejected it as being too costly and laborious. Hudson encountered the same opposition by farmers in his Utah study.[1]

Nevertheless, there are some ambitious farmers who have included this option in their irrigation system. In La Joya Antigua, one farmer built his house adjacent to the principal canal but above its natural command area. He acquired a pump to bring water to the construction site and eventually used this pumping system to irrigate a hectare of fruit trees surrounding his new home (figure 28).

Another farmer raised water beyond the area commanded by the principal ditch near the town of Uchumayo just downstream from the Campiña along the Chili River. He had acquired an additional 3 ha of arid land through a petition to the national government. This new land received water that was pumped from a lower lying irrigation ditch to an elevation of about 40 m, where it was collected in a small concrete tank readied for gravity irrigation. The cost of this installation was several thousand soles, but the farmer was not too concerned as he believed he would recoup his initial investment in 5 years. At the time of the field survey he had planted alfalfa to enrich this land and planned later to set out fruit trees. A neighbor, having seen the effectiveness of this pumping system, immediately petitioned the national government for 10 ha of arid land above his main irrigation canal in order also to use more of his water quota.

In a similar manner, a difficult and costly undertaking in raising water above the area commanded by a ditch by pumping occurred in the Chilina zone within the Campiña. In this instance a farmer planned to pump water 60 m in two stages with a holding tank at 30 m and another at the top of the ascent. The amount of land to be irrigated was 7 ha. In actuality due to the large expenditure required, the completion of this project was in doubt.

Various motives underlie the actions taken by these farmers. The La Joya farmer's actions, not initially related to agriculture but to the construction of a home, reflect a set of noncommitted sequential events. On the other hand, the Uchumayo farmers, who were not using their full water rights when they decided to pump water, made a decision at the juncture of a chain reaction founded on a previous commitment to use their full quota of water. If given similar opportunities, other farmers would surely initiate a similar operation in order to gain full use of their water rights. Normally

---

[1] Ibid., p. 162.

Fig. 28. Pump installed along the main canal in La Joya Antigua. This pump later was used to lift water above the ditch to irrigate additional land not part of the original farm.

these opportunities are found where water is abundant, as near the head of the *Campiña* in the Chilina area, or just downstream from the *Campiña* near Uchumayo.

Most farmers in the study area use concrete headgates to divert water from the community canal to the farm. In the Chichas area and in most of the *Campiña*, farm headgates originally had a well constructed metallic gate. Many of these are now in disarray with the gates missing, or so battered at the lower end that water continues to flow into the farm ditch after the gates are closed. To correct this problem, the government developed a plan to replace missing or poorly functioning gates, only to have this program strongly opposed by farmers who had grown accustomed to a seemingly adequate system of diverting water. Generally in La Joya Antigua and universally in San Isidro, conditions of farm headgates are different from those in the *Campiña*. On the Pampa of La Joya the better farm headgates include both a well functioning metallic gate and a locking system.

In order to facilitate the diversion of water through a farm headgate, several techniques exist in the study area for temporarily elevating water in the community ditch. Throughout the Chichas area and in several places in La Joya Antigua, large rocks and blocks of sod (*Chamba*), placed just downstream from the farm headgate within the community ditch, act as a check to control the elevation of the water surface. *Chamba*, consisting mostly of Kikuyo grass (*Pennisetum clandestinum*), grows along ditch borders. Thus, this grass is readily accessible for use by irrigators for sod. In time, however, removal of too much *Chamba* can weaken a border. Kikuyo grass, growing along a ditch, is a useful resource, but in an alfalfa field, it is a bothersome weed, which results in a diminution in the length of time an alfalfa field can exist.

Another manner of elevating the water surface in community canals is through the use of portable wooden checks set within iron grooves along the walls of canals, located just downstream from the farm headgate. Many farmers in La Joya Antigua, and all in San Isidro, use this technique.

Land preparation. The leveling of land is an integral part of most types of gravity irrigation. Both border strip and furrow irrigation methods operate more effectively when the land has been properly leveled. In the Chili Basin, most farmers preparing a field for a new crop attempt to re-level the land with a crude wooden board drawn across the field with a pair of oxen. This operation occurs even on the small terraced fields found in the

Campiña. Farmers realize that their leveling may be inadequate but few really have anyone to turn to for additional help in earth moving and establishing levels. Where the land is shallow and stony, farmers are reluctant to do too much leveling. Land tenureship does not seem to affect this farm operation; both renters as well as owner-operators make an attempt to level their land before planting.

In San Isidro, where new land was being put under the ditch for the first time, the initial leveling was not entirely successful. Although the technical staff of the government with the aid of such sophisticated equipment as tractors and scrapers worked with the farmers in leveling the land for border strip irrigation, the first waterings did not advance down the strips uniformally. Corrective action taken by farmers in reshaping the lateral slope during these initial waterings led to some improvement.

Similar to what Hudson found in the Utah Valley, farmers in the Chili Basin are hampered from doing a proper job of leveling.[1] In part, this is due to the nature of the physical environment, the lack of capital, proper technical aid, and more sophisticated equipment. Attempts to cope with this task are not easy.

The application of water to fields. Similar to diverting water from community ditches to farms, irrigators use gravity to get water from farm ditches to fields. In both the Campiña and La Joya Antigua, water flows from the farm ditch to the field through gaps (boquetes) broken by shovel in the banks of farm ditches.

The number and placement of these openings depend on the size, shape, and lay of the field; the cost of construction; and the method of irrigation. While fields under one ha generally have one opening, those larger than this might have two or more, as shown in figure 29 for the Chichas area. But due largely to more regularly shaped fields, there exist fewer openings to fields from farm ditches in La Joya Antigua than in the Campiña and in particular the Chichas area.

To prepare for an orderly watering during the next turn, farmers may close these openings after a field receives water. If this procedure is not followed, at the beginning of the next watering a field that did not have these openings closed might unintentionally receive water while the irrigator was watering

---

[1]Ibid., p. 161.

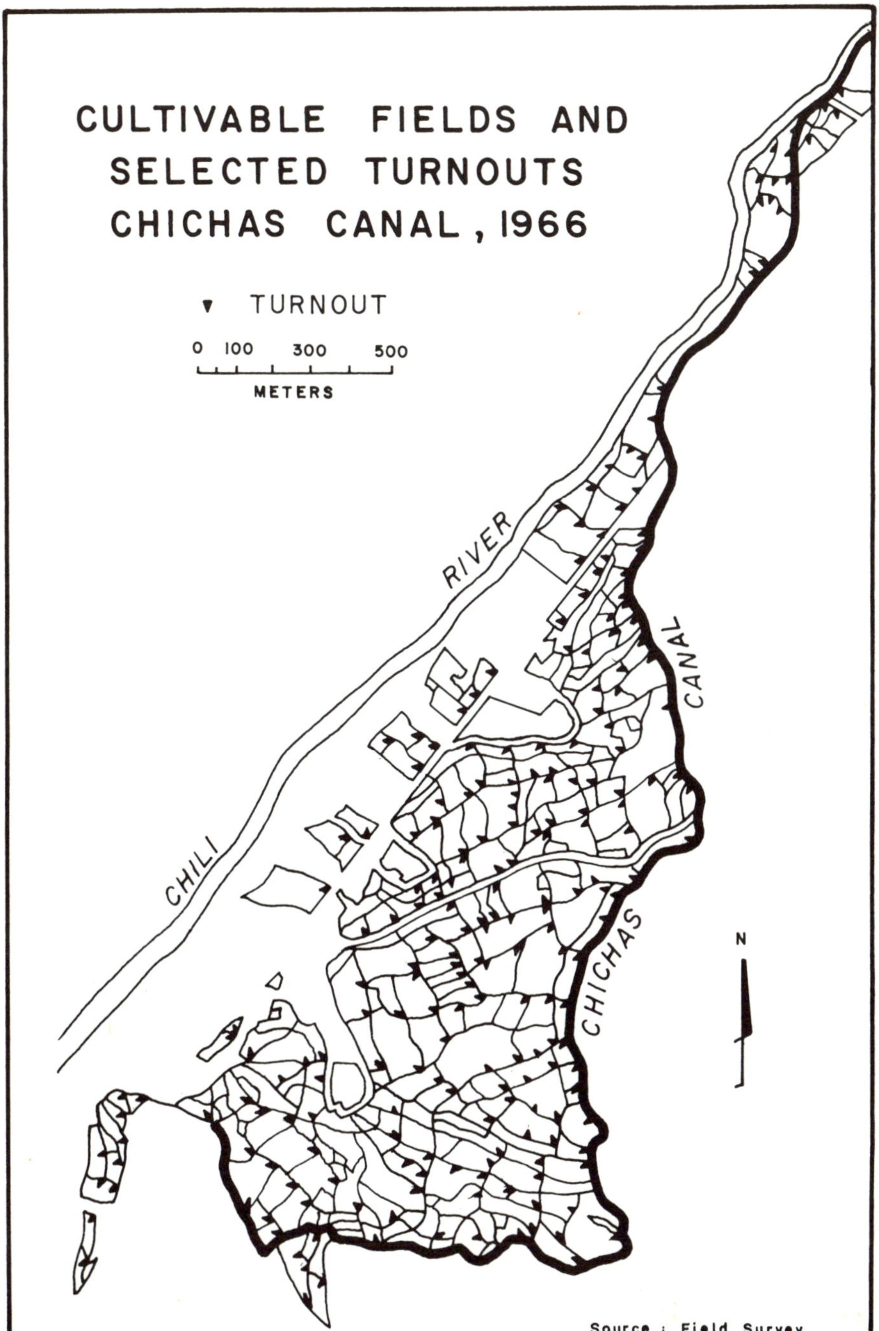

Fig. 29 - Cultivable Fields and Selected Turnouts, Chichas Canal, 1966.

another field according to his itinerary. Irrigators close these gaps with rock and Chamba (figures 30 and 31).

Although water generally flows to fields from farm ditches in the Campiña and in La Joya Antigua, water in San Isidro arrives at the field by way of a community feeder canal connected to a lateral. This feeder canal may supply water to two or three farms. Along the cement-lined feeder canal permanent openings exist, allowing water to flow directly into a farmer's head ditch. Indeed, for each head ditch there is an opening to the feeder canal.

For watering their fields, farmers in the Chili Basin use several irrigation methods based on gravity flow of water. The most widely employed methods are furrow, border strip, and wild flooding. The degree of control over irrigation water is greatest with furrow and border strip irrigation. Summary field data of ditching used for furrow and border strip irrigation as found in the study area appear in table 29. In describing and analyzing these methods, the distribution system used as to ditching and the routing of water is outlined.

Furrow and border strip. Just as Rawitz indicates ". . . that there is no difference in principle between border and furrow irrigation . . .,"[1] so the nature of the two methods will be described jointly. Water, on entering a field, immediately flows into a head ditch (acequia regadora) constructed along the high edge of a field (figures 32 and 33). Although there usually exists only one head ditch for each field, at times farmers place a second head ditch midway down the field. This additional head ditch occurs when individually or in combination the length of the field seems excessive, the longitudinal slope is great, or the soils are shallow. To bring water from the head of the field to the head ditch in mid-field, farmers use a lateral ditch placed perpendicular to the head ditches and along the border of the field. An example where excessive slope and shallow rocky soil necessitates the use of more than one head ditch in a field occurs in the Zamácola Irrigation Project in the Campiña. Throughout the study area the slope of the head ditch is generally less than 1 percent though sometimes it is more than 2 percent. Again, the Zamácola Irrigation Project contains large gradients along head ditches. Several of these ditches with an excessive slope lie perpendicular to the contour of the land.

---

[1]E. Rawitz, "Gravity Irrigation," in Arid Zone Irrigation, ed. B. Yaron, E. Danfors, and Y. Vaadia (New York: Springer-Verlag, 1973), p. 327.

Fig. 30. Closing an opening along a farm ditch in the Chichas area. Irrigator places rock and sod in the opening, upstream view.

Fig. 31. Continuing the closing of an opening along a farm ditch in the Chichas area, downstream view (See figure 30). The sod between the rocks creates a relatively tight fitting seal.

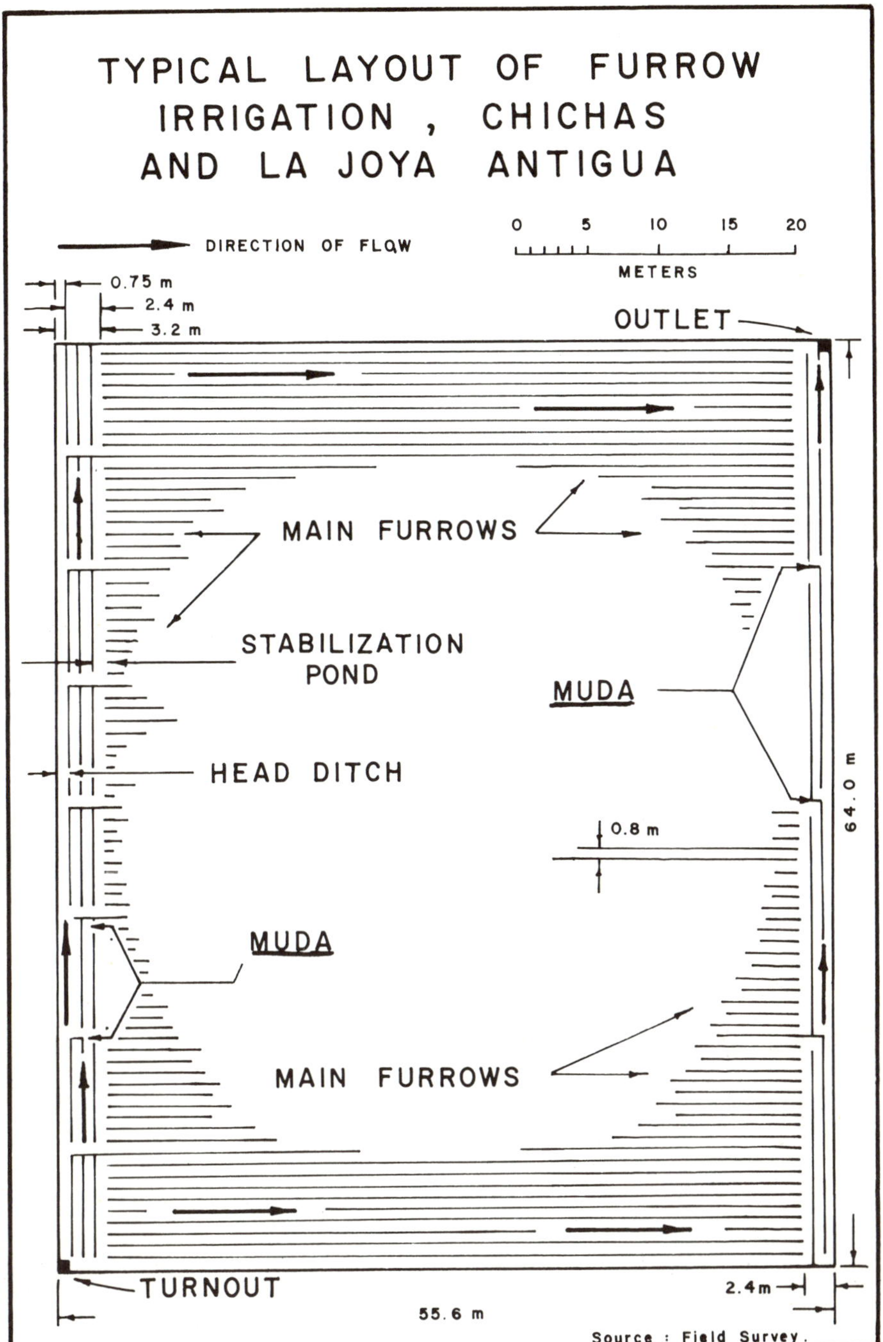

Fig. 32 - Typical Layout of Furrow Irrigation , Chichas and La Joya Antigua .

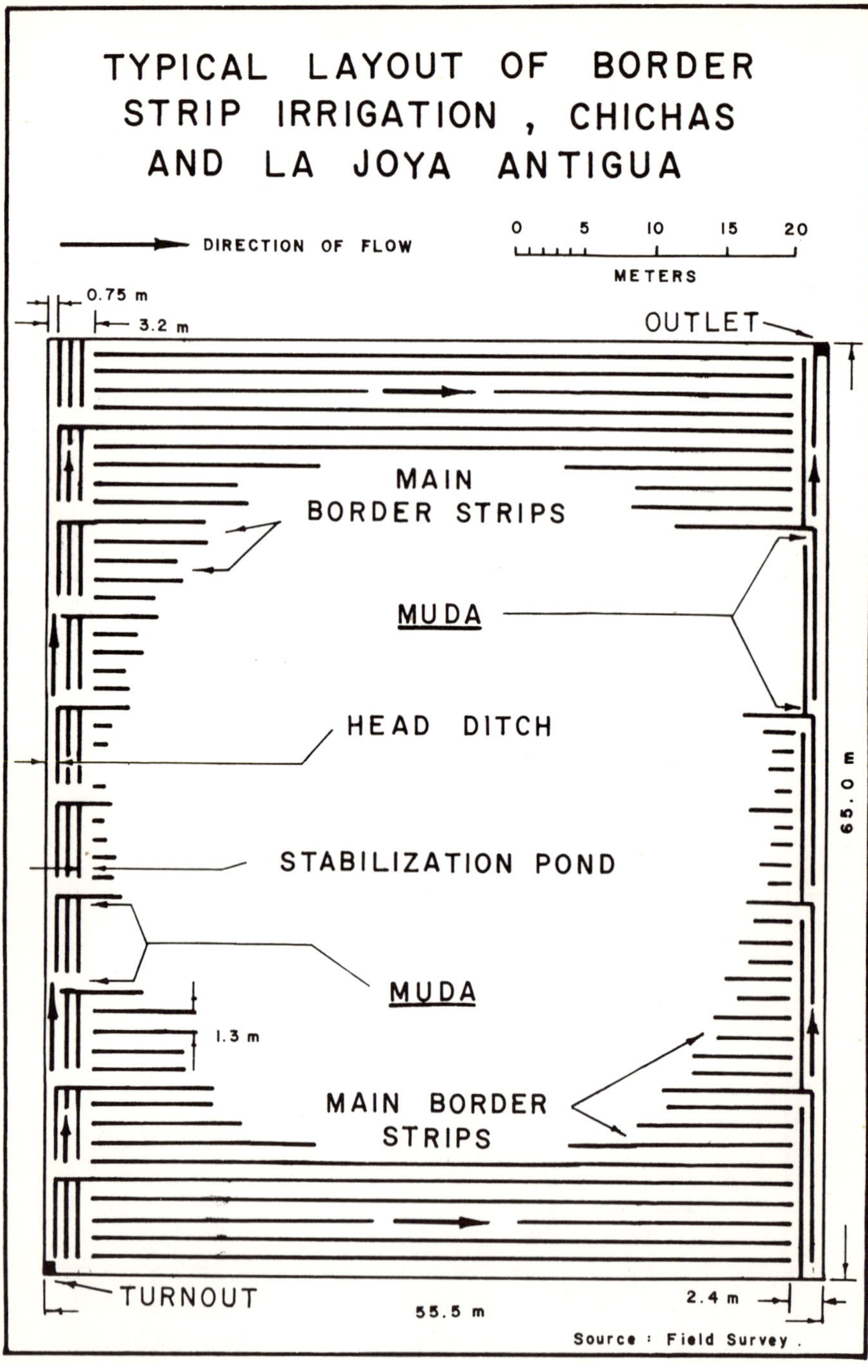

Fig. 33 - Typical Layout of Border Strip Irrigation , Chichas and La Joya Antigua.

TABLE 29

NATURE OF CONVEYANCE SYSTEM OF FURROW AND BORDER STRIP IRRIGATION CAMPIÑA OF AREQUIPA AND LA JOYA ANTIGUA, 1966

Zone and Method of Irrigation	Number of Fields	Type of Conveyance		Average Slope (%)	Average Length (m)
		Type[a]	Number Measured		
Chichas Furrow Irrigation	7	HD	18	0.85	59.43
		L	4	0.84	51.38
		HF	66	1.22	7.72
		MF	71	0.98	41.95
		TF	12	0.88	60.61
Chichas Border Strip Irrigation	2	HD	2	0.51	102.45
		MBS	9	1.13	48.99
		TBS	1	0.31	112.10
Zamácola Furrow Irrigation	3	HD	11	2.35	54.33
		L	3	1.48	75.00
		MF	32	1.31	40.65
La Joya Antigua Furrow Irrigation	5	HD	11	1.56	96.42
		L	1	0.44	330.00
		MF	38	1.27	61.93
		TF	2	1.88	41.50
La Joya Antigua Border Strip Irrigation	3	HD	7	0.67	29.95
		MBS	6	1.51	32.27[b]

SOURCE: Field Survey. Due to limited field measurements, text also considers general observations and local technicians' views.

[a]HD=Head ditch; L=Lateral ditch; HF=Head furrow; MF=Main furrow; TF=Terminal furrow; MBS=Main border strip; TBS=Terminal border strip.

[b]This measurement comes from only one field.

After water enters the head ditch, the water flows through a series of openings (boquetes) cut by shovel in the earth border of the head ditch; the distance between boquetes varies from 4 to 20 m, depending upon the design of the irrigation method. These openings allow water to flow into a set of relatively short head furrows or border strips (mudas) which parallel the head ditch. For row crops such as corn and vegetables, furrows are used; whereas, for small grains and alfalfa border strips are employed. Normally, 3 to 4 head furrows constitute a muda or 2 to 3 head border strips constitute a muda, depending upon the crop. These mudas vary in length from 4 to 20 m, the average being 8 m.

Perpendicular to these mudas, the main furrows or border strips occupy the bulk of the field. Water flows from the last furrow or border strip of the muda into the main furrows or strips. Thus, this last trench functions somewhat as a stabilization pond. However, for this last trench to act as a true stabilization pond, the level of water in it would have to be at the same elevation as the water entering the opening to the main furrow or border strip to permit the water to flow down field at a fairly constant rate.[1] This condition of forming a stabilization pond, though, does not always occur in the study area.

Due to imperfections in the design of these mudas and the unexpected fluctuations in the rate of water flow received at the field, irrigators, in order to effectively utilize this special layout, attempt to control the flow of water at the head of the field by allowing water to enter one or more muda simultaneously, according to the amount of water available. In controlling the flow of water, the irrigator tries to avoid having the water erode the gaps in the mudas while allowing a sufficient amount of water to flow down field. To aid in decreasing the erosion of these gaps, an irrigator also lays Chamba against the sides of the openings.

A positive function of the muda at the head of a field is to keep erosion of plants and soil to a minimum, especially before alfalfa has established itself in border strips. Nevertheless, since mudas were not initially employed in San Isidro, a serious problem of erosion occurred at the head of these strips as designed by the technician.

Generally, the length of the main furrows and border strips are constant throughout a field, except where the field is irregularly shaped. The length of these furrows and border strips fluctuates between 40 and 50 m with a slight tendency for them to be about 10 m longer in La Joya Antigua than in the Chichas area. Since the soils in the Campiña and especially in the Chichas area are heavier than on the Pampa of La Joya, the length of both furrows and border strips should normally be longer in the former areas, but this generally does not occur. Short furrows and border strips in the Chichas area naturally exist where fields are small. Even where a run could be 80 m or more in length, farmers normally divide the field in half by placing a head ditch in mid-field to adjust for steep gradients. Therefore, since furrows and border strips in the Chichas area are not long, a large amount of land is

---

[1] Houk, Irrigation Engineering, I, 510.

lost in borders and ditches. It appears that custom and small fields necessitated by rough topography as especially found in the Campiña outweigh the increased labor cost and loss of productive land incurred in resorting to relatively short runs. As Booher indicates,

> Care is needed in selecting the proper length of furrow to use. Labor requirements and costs for irrigating increase as the furrows become shorter; while the uniformity of water application usually decreases as the furrows become longer. The furrows should, therefore, be as long as will permit reasonable efficiency in the application of the water.[1]

On the other hand, in meeting the goal of standardization of water management throughout San Isidro, engineers routinely laid out border strips 60 m in length which determined the spacing between the cement-lined head ditches (figure 34). As Rawitz states,

> . . . it is not worthwhile to have runs shorter than about 60 m; if such short runs appear necessary, it is an indication of very high infiltration capacity, and a switch to sprinkling should be considered.[2]

In fact, sprinkler irrigation is programed for the newer extensions of the overall La Joya Irrigation Project.[3]

Another critical parameter besides the length of runs of furrows and border strip irrigation is slope. In the established irrigated areas, the average slope of the main furrows and border strips fluctuates between 1.0 and 1.5 percent, with some situations though as high as 2.5 percent. In general no notable difference exists in slope between the fields of the Chichas area and those of La Joya Antigua. But where the gradients were great an appreciable amount of excess soil lay at the foot of fields. Both this observation of excess erosion, as noted in the Campiña, and the comment of Booher that, "To avoid excessive erosion the slope of the furrows in the direction of water movement should not exceed 2 percent when irrigating cultivated crops,"[4] lead one to conclude that farmers are probably using runs with too great a longitudinal slope for furrow irrigation.

---

[1] Booher, Surface Irrigation, p. 128.

[2] Rawitz, "Gravity Irrigation," p. 332.

[3] Equipo Agropecuario Peruano-Israel, Proyecto de Planificación, Colonización "La Joya."

[4] Booher, Surface Irrigation, p. 122.

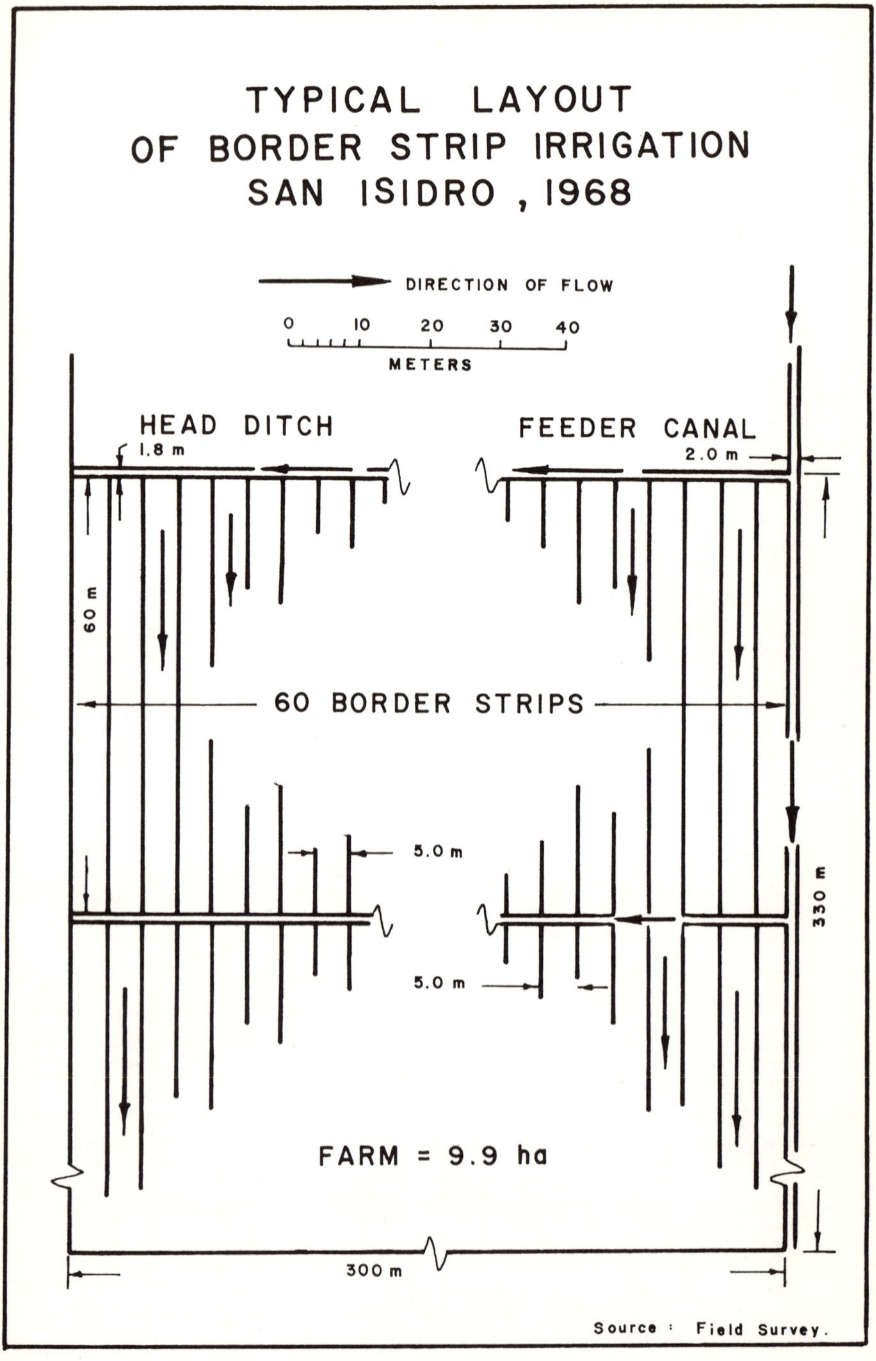

Fig. 34 - Typical Layout of Border Strip Irrigation, San Isidro, 1968.

In San Isidro, however, slopes of border strips are gentle, ranging from 0.2 to 0.3 percent. This range in slope is in accord with the thinking of Booher, who states that,

> A minimum slope of 0.2 percent is recommended when irrigating lucerne (alfalfa), orchards or vineyards, which require relatively infrequent heavy irrigations because of their deep rooting characteristics.[1]

Yet Zimmerman states that, "Within limits of 0.4 to 1.5 percent the slope can rarely be too steep for forage crops."[2] With alfalfa planted in border strips, the problem with steep slopes occurs during the period when the crop is being established.

The width of furrows and of border strips also characterizes these two irrigation methods. Factors which affect this distance are the type of soil, the manner of cultivating the soil, and the method of applying water. For example, all other factors remaining constant, the lighter the soil texture, the closer together the placement of furrows. As Doneen indicates, "(If) furrows are placed too far apart, portions of the soil between them will remain dry, especially when the soil is sandy . . ."[3] Therefore, in considering this one variable, one might expect the distance between furrows for equivalent crops to be greater in the Campiña than in the Pampa of La Joya irrigation projects. This condition may exist, but due to the lack of sufficient data this assumption could not be empirically verified. Neither could any appreciable difference in width between furrows be detected within an area. Based on the few measurements made, the width of furrows in the study area fluctuates between 60 and 80 cm.

The width of border strips in the study area is approximately 1 m except for those in San Isidro where the engineers designed them to range from 3 to 6 m and in several instances up to 10 m. But traditionally farmers in the established areas of the Campiña and La Joya Antigua, by using narrow border strips, can adjust readily to great fluctuations in the flow of water received at the farm headgate. With a small flow it is more feasible for a farmer to irrigate a narrow border strip than a wide one. With a small flow, a farmer finds it difficult to flood a wide border

---

[1] Booher, Surface Irrigation, p. 101.

[2] Zimmerman, Irrigation, p. 134.

[3] L.D. Doneen, Irrigation Practice and Water Management, Irrigation and Drainage Paper no. 1 (Rome: Food and Agriculture Organization of the United Nations, 1971), p. 31.

strip because it does not result in a uniform advancement of water down field.

Also, there exists another advantage in using narrow border strips. As previously indicated, farmers cannot properly level their fields. Yet, by using narrow border strips farmers can overcome, in part, difficulties resulting from unevenness of the land surface.[1]

On the other hand, in San Isidro engineers, contemplating a fixed flow of 100 l/s for each farm during the colonization period, believed that border strips with a width 3 m or greater would be appropriate. This expectation was not realized as described more fully later.

Finally, on reaching the end of the main furrows or border strips, water flows into a set of mudas similar in layout to those at the head of the field. These terminal furrows or border strips allow the water at the foot of the field to be more evenly distributed and also to percolate more fully into the soil before flowing off field. These terminal mudas are, however, only present in the established irrigated areas. In San Isidro they do not occur; here the excess water collects at the lower end of the strip without any outlet. The lower end of these strips abut the adjoining head ditch used for supplying water to the next set of border strips further downstream.

Wild flooding. Farmers use border strips in irrigating alfalfa only during its early growth period, before grazing. After the first grazing when the borders are trampled upon and destroyed, the original irrigation method becomes useless and wild flooding is adopted.

Under wild flooding water enters the field through openings in the head ditch; water then is allowed to flow freely down the slope. In many instances, farmers construct temporary interceptor ditches leading from these openings and place them across the slope allowing the water to overflow the banks. Sometimes openings are made in the interceptor ditches to direct the flow of water down slope.

Several major problems occur in using wild flooding. As Booher states,

---

[1] Booher, Surface Irrigation, p. 102.

> The success of this method of irrigation is often dependent upon the skill used in locating the points where water is released from the . . . ditches and in adjusting the size of the openings so that the correct amount of water is released to cover the area served without causing soil erosion.[1]

In the established irrigated areas in the Campiña and in La Joya Antigua, farmers use the already existing large openings in the head ditches for wild flooding. By this time alfalfa has been established and there is less of an opportunity for erosion to take place. But, in San Isidro a problem developed in getting water from the head ditch to the field when wild flooding was unexpectedly adopted. Consequently, before this type of irrigation can be used effectively, considerable experience is needed.

Also, the attention and skill of the irrigator are important for the successful application of water by wild flooding. Even though some land leveling takes place in preparation for border strip irrigation, the irrigator must constantly observe the flow of water and must change settings frequently in order to successfully apply water. These attributes of the irrigator are especially needed when fields are small. Also, at night when many irrigators water alfalfa by wild flooding, the irrigator's attention and skill may become somewhat lax and result in poor application of water.

Another problem confronting the farmer in using wild flooding is the risk of increased losses of water by surface runoff and deep percolation while attempting to avoid the cost of maintaining border strips. Again, this problem is most serious during night waterings.

Summary. Largely due to the influence of tradition, farmers rely upon similar layouts for a particular irrigation method throughout the established irrigated areas. This commonality in design also exists in San Isidro once border strip irrigation gives way to wild flooding.

Nevertheless, since farmers have only a limited ability to modify the environment, notable differences do occur in the design of a particular irrigation method between and within zones. Some

---

[1] Ibid., p. 115.

of the variables responsible for these differences include the size and shape of a field; the nature of the soil, especially the texture and depth; the perceived availability of water, including its fluctuation; and the lay of the land. For example, within the Chichas area a wide range in size and shape of fields exists which in part influences the length and gradient of irrigation runs. On the other hand, zonal variations occur with border strip irrigation whose design is markedly different between San Isidro and the other established irrigated areas. Factors which do not seem to play a critical role in affecting the design of an irrigation method are the manner of delivering water and the farmer's age, economic standing and land tenure status.

Throughout the study area, the decisions made by resource managers in designing a particular irrigation method occur under conditions of commitment. In the established irrigated areas, generally the personal preferences of farmers lead to this form of decision-making. On the other hand, in San Isidro, engineers made decisions which in part reflect the importance of their experience obtained when they participated in the development of the San Lorenzo Irrigation Project in northern Peru.

In comparing the differences in design of irrigation methods between those developed by the farmer and those by the engineer, a marked variance exists. It appears farmers have been flexible in their handling of the important variables of topography, land leveling, and stream flow. For example, farmers compensate for poor leveling by instituting narrow border strips which in turn make for greater accommodation to the wide fluctuations in the stream flow received at a field. To be sure, engineers, who are cognizant of these important factors, try to cope with them instead by relying upon past experience and standard engineering manuals.

Just as Hudson observed for the Utah Valley, farmers of the Chili Basin in laying out an irrigation method, have had to contend with an array of variables, many interrelated.[1] They met this challenge on their own without the aid of the expert and have developed a set of irrigation methods that has lasted unchanged for years. Now, with an ever increasing demand on the limited hydraulic resources of the Chili Basin, the efficacy of these traditional irrigation methods will be questioned and new ones will be tried, some successfully. In this rush toward seeking a more

---

[1] Hudson, Irrigation Water Use in the Utah Valley, Utah, p. 167.

effective means of applying water, it is well to remember that Houk states,

> In some regions farmers have become accustomed to particular methods of applying water and continue to use such methods when others might be more desirable and economic.[1]

Scheduling of Waterings

Following Hudson's findings, important elements in the scheduling of irrigations are the factors that influence the timing of a watering, the length of the irrigation season, and the number of irrigations.[2]

Factors affecting scheduling. From a scientific and practical stance, "Two major considerations influence the time of irrigation; namely, (a) the water needs of the crops, and (b) the availability of water with which to irrigate." [3] After many discussions with farmers in the study area while observing their irrigation practices, specific factors that affect the scheduling of water were noted and the range of alternatives considered by farmers were found to be more limited than those of the technical-scientific community.

1. Water delivery. The interval between the delivery of water to farmers in the study area varies greatly depending upon the location of the farm. This interval can range from 3 to 16 days. In the past, before the institution of the new 1969 Water Code, some farmers had access to water on an almost daily basis. In general, though, the rotation system of delivery, as evidenced in the study area, presents farmers with two major problems. When the interval between turns is a week or more, farmers who need water more frequently during the critical first weeks of a newly planted alfalfa field, face the task of obtaining extralegal water, either by borrowing or using other means. Also, a farmer may not need all or only a portion of a turn (mita) when there has been a recent rain or when much of the farm is in fallow or when crops are being harvested. Even so, at these times a farmer cannot afford to use

---

[1] Houk, Irrigation Engineering, 1, 501.

[2] Hudson, Immigration Water Use in the Utah Valley, Utah, p. 154.

[3] Orson W. Israelsen, Irrigation Principles and Practices, 2d ed., (New York: John Wiley and Sons, Inc., 1950), p. 284.

all of his mita. If he were to skip a mita, it might be too long an interval between irrigations. As a result, by using a mita under these conditions, gross waste of water may occur. As many experts including Haise and Hagan concur, "Maximum flexibility and the greatest opportunity to irrigate on a more scientific basis occurs where irrigation water is available on demand."[1] Until recently the demand system would have been impossible along the Chili River, yet some farmers with small farm ponds practice a modified form of it. Now with the new large dams and interbasin transfers, the opportunity exists for officials to institute a more optimal water delivery system. This would entail a great change and would probably clash with the existing traditions of farmers and possibly the local technicians managing the water resources and the irrigation schemes.

Upon receiving a mita, a farmer must decide which fields to irrigate. Previous to the functioning of the new large dams and interbasin transfers, farmers in the Chichas area during high stream flow generally had an abundance of water. At this time of the year, farmers were able to irrigate fields they wanted to during a mita. This optimal condition though did not usually exist during the low stream flow season. Since the Chichas Canal was the last downstream canal in the Campiña to receive water from the Chili River on a percentage basis, the farmers faced a situation whereby fields which needed water during a mita might be skipped and some crops, especially alfalfa, left in a droughty condition. On the other hand, in La Joya Antigua before the operation of the large dams and interbasin transfers, some farmers did not have sufficient water throughout the year to plant all their land under the ditch. Now with the institution of a regulated stream flow there is more water available to all on a secure basis, and the possibility exists for a more relaxed situation which would allow farmers to make water scheduling decisions more in line with the water needs of the crops in particular and farm operations in general.

2. The permeability of soils. On the Pampa of La Joya where the soils have a high permeability rate, farmers believe that water should be applied frequently and heavily. In the Campiña and es-

---

[1] Howard R. Haise and Robert M. Hagan, "Soil, Plant, and Evaporative Measurements as Criteria for Scheduling Irrigation," in Irrigation of Agricultural Lands, ed. Robert M. Hagan et al. (Madison, Wisconsin: American Society of Agronomy Publications, 1967), p. 598.

pecially in the Chichas area where the soils contain appreciable amounts of silt and clay, farmers believe less frequent waterings are sufficient. But, similar to farmers on the Pampa, they, too, believe applications should be heavy. With good natural drainage in both areas, excess field water has an opportunity to flow away leaving most fields well aerated.

3. Plant water indicators. Just as Hudson observed among farmers in his Utah study, so, too, do most farmers in the Chili Basin judge the need for water by the appearance of the crop.[1] A change in the color of the leaves of certain crops announces a critical period. For example, dark green alfalfa signals to the farmer that the plant suffers from a lack of water. Also the wide use of leaf wilting guides farmers in the scheduling of water. Farmers rely in part on this sign in watering corn. That these are practical guides in scheduling waterings is evident, but if relied on wholly, a farmer might run the risk of having waited too long in watering a crop. By this time sufficient damage could have occurred to negatively affect the final yield. Indeed, according to Israelsen and Hansen, "Crop growth should not be retarded by lack of available soil moisture; and the practice of withholding irrigation until the crop definitely shows a need for water is likely to retard growth."[2]

Several other visual indicators of water stress exist but none are used by farmers of the area. These indicators include: plant movement, exudation, and plants selected and managed for use as indicators for irrigation.[3] In the near future plant indicators might possibly be introduced into the area. For example, the existing practice of double-cropping of cabbage with other vegetables might present an opportune situation for the introduction of this technique that meshes with existing traditions.

Other methods available for detecting water stress in plants include such plant growth indicators as leaf reflectance and temperature, and plant water measurements.[4] For proper use,

---

[1] Hudson, Irrigation Water Use in the Utah Valley, Utah, p. 154.

[2] Israelsen and Hansen, Irrigation Principles and Practices, p. 268.

[3] Haise and Hagan, "Soil, Plant, and Evaporative Measurements as Criteria for Scheduling Irrigation," pp. 587-588.

[4] Ibid., pp. 588-592 and H. Bielorai, "Prediction of Irrigation Needs," in Arid Zone Irrigation, pp. 364-365.

though, these techniques need a high degree of sophisticated technical input. As a result, at the present time they cannot be considered as part of the array of alternatives which farmers can consider in scheduling waterings. As Bielorai says, "These measurements must be carried out by skilled personnel, and are thus impractical for routine use by farmers." [1]

4. Stages in the vegetative cycle. Depending upon the growth stage of the crop, farmers schedule waterings apparently related to the crop's water needs. Stages, when farmers are especially conscious about properly scheduling waterings, occur during the period when alfalfa is being established and when onions are first set out from the seedbed. Other times when farmers give a high priority in scheduling irrigations occur during the formation of the grain of cereals, the tubers of potatoes, and the bulbs of onions and garlic. In addition, once onion and garlic bulbs have fully developed the scheduling of irrigation is prudently reduced to avoid the putrification and second growth of the bulb.

Other traditions among farmers have evolved which affect the scheduling of irrigations not related to crop water needs but to cultural practices tied to the vegetative cycle of the crop. Some scheduling takes place in conjunction with the harvest season. To facilitate the gathering of onions and garlic farmers schedule an irrigation just previous to the harvest in order to loosen the soil. This practice is common in the Chichas area where the soils are heavy. Also, in the harvesting of onions, when the market price is deemed too low, farmers may withhold the onions temporarily from the market by keeping them in the ground 3 or 4 weeks beyond the normal harvest time, postponing the final irrigation until just before the harvest. By this action farmers hope that during this time the price of onions increases.

Another crop in which harvesting affects the scheduling of irrigation is alfalfa. Customarily, farmers irrigate alfalfa just before allowing livestock to graze it. They do not normally water the crop during the grazing period. Fundamentally this scheduling procedure is in accord with the experts' views. Indeed, Keller and Carlson state that,

> If the soil has become quite dry, irrigating several days

---

[1] Ibid., p. 364.

> before harvest may get the next crop started more quickly rather than delaying irrigation until the current crop has been removed.[1]

And they go on to add that, ". . . livestock grazing upon pastures still wet from irrigation compacts the soil and reduces the rate of water intake."[2] Nevertheless, some irrigation of alfalfa being grazed does take place in La Joya Antigua but only a little in the Chichas area. Realizing that the former area contains lighter soils than the latter area, one can appreciate the difference between the two. An additional cultural practice, not completely related to plant water needs determines the scheduling of irrigation water for alfalfa. In aiding the process of germinating alfalfa seeds in a field coated with a thin layer of silt, a farmer may schedule an additional irrigation to soften the silt and save the crop.

Another custom that seems to have some scientific basis is the scheduling of an extra watering for young corn, not so much to meet the water needs of the crop, although this is in part accomplished, but to kill a worm that begins to attack the roots and stalk of the tender crop. By applying water at this early period, literally farmers hope to drown the insect, thus the name of this irrigation: matagusano.

5. The relative value of different crops. When the supply of water is limited, farmers, in scheduling irrigations, give preference to the cash crops of garlic and onions over all other crops. Next in importance are potatoes, corn, small grains, and finally, alfalfa.

6. The appearance of the soil. Scientists have prepared a simple guide for judging the amount of available moisture for crops by feeling and looking at the texture of the soil.[3] Yet no farmer in the study area inspects the soil profile with a shovel in order to determine the moisture content and the time to irrigate. Admittedly, some scientists in evaluating soil moisture believe that a more useful tool for scheduling irrigations is the use of soil-

---

[1] Wesley Keller and Carl W. Carlson, "Forage Crops," in Irrigation of Agricultural Lands, p. 612.

[2] Ibid., p. 613.

[3] U.S. Soil Conservation Service and U.S. Bureau of Reclamation, Irrigation on Western Farms, Agriculture Information Bulletin No. 199 (Washington D.C.: Government Printing Office, 1959), pp. 42-44.

water potential or suction.[1] The use of this technique, however, would entail much more sophisticated input in both equipment and trained personnel than is available now or in the foreseeable future.

Irrigation scheduling in the Chili Basin. Much of what has been cited as the factors influencing the scheduling of waterings are substantiated by the data gathered from the field survey. For at least 1.5 years during 1965, 1966, and 1967 several farms in the study area were constantly inspected with the aid of various field helpers, one of whom lived in La Joya Antigua. This particular inquiry on irrigation scheduling includes 14 farms in La Joya Antigua, 2 farms in the Chichas area, and 1 farm in the Zamácola Irrigation Project. Some of the original data had to be discarded due to a lack of continuity in field observations and measurements brought about in large part by unanticipated waterings by irrigators.

The semiautonomous status of the camayo, who may act as the administrator of a farm or at least as the chief irrigator, holds a key role in irrigation, especially when the absentee owner or renter allows him to schedule and supervise the irrigation process. In many instances the camayo who often comes from the high sierra has a tendency to practice a form of horticulture whereby, in scheduling irrigations, he will consider each field as entirely unique from all others. With this in mind, information for each field is tabulated independently of all others and not combined according to crop type to form "irrigation situations" as Hudson did in his study.[2]

Irrigation season. Since crops can be grown year-round in the Chili Basin, the irrigation season continues throughout the year. Before the regularization of the stream flow of the Chili River, farmers, during September and October, when the stream flow became exceedingly low, would have to drastically curtail the scheduling of waterings. At this time many fields would lie in fallow waiting to be irrigated with an augmented steam flow usually arriving in December. Newly planted corn benefits from these

---

[1] Bielorai, "Prediction of Irrigation Needs," pp. 362-363.

[2] Hudson, Irrigation Water Use in the Utah Valley, Utah, p. 155.

scheduled waterings at the beginning of the hydrologic year. In planting corn at this time of the year, farmers can harvest the crop 6 months later avoiding the occasional early frosts occurring between June and August in the Campiña. The scheduling of waterings for this crop begins to increase in December, reaching a peak in March and April, and tapers off in May.

The only discernible difference in the time of planting and harvesting corn among farmers within a zone occurs in the Chichas area. Farmers at the head of the canal who have access to water on a daily basis, plant corn 2 or 3 weeks earlier than the others. Comparing the vegetative cycle of corn between Chichas and La Joya Antigua, the cycle for the latter area is about two weeks shorter, giving a slightly shorter irrigation season for this crop on the Pampa.

Similar to corn, farmers plant potatoes near the beginning of the hydrologic year and harvest them 6 or 7 months later. For onions and garlic farmers plant them during the middle of the hydrologic year and harvest them 5 to 8 months later.

Alfalfa, another important crop of the Chili Basin, grows year-round. Depending on its care and treatment, alfalfa can last from 4 to 6 years. An important factor determining this crop's longevity depends on the rate that Kikuyo grass overtakes it. Readily available water being crucial to alfalfa during the first month after planting, farmers sow it between January and June. The irrigation season for alfalfa does not vary greatly either within or among zones.

Sometimes in association with alfalfa, farmers sow either barley or wheat to be harvested before the first "cutting" of alfalfa. When planted alone, both barley and wheat have similar irrigation seasons. They are planted in mid-year and harvested in January, giving a 6 month irrigation season.

The irrigation season can also be analyzed in terms of the number of months waterings are scheduled (table 30). In general, the crops grown in the area can be divided into 5 groups according to the number of months scheduling of water occurs. Small grains and potatoes are irrigated from 3 to 5 months; corn, 2 to 6 months; onions, 4 to 8 months; other vegetables, 7 to 8 months; and alfalfa for 12 months consecutively. The 3 corn fields watered for only 2 to 3 months reflect a short-lived crop where farmers allowed cattle to graze the tender crop.

In conclusion, there appears to exist a slight tendency for

TABLE 30

NUMBER OF MONTHS IN WHICH FIELDS WERE IRRIGATED, CHILI BASIN, 1965-1967

Crop Type and Location	Number of Fields								Total
	Number of Months								
	1	2	3	4	5	6	7	8	
Corn									
La Joya Antigua		2	1	15	24	9	-	-	51
Chichas		-	-	5	4	-	-	-	9
Zamácola		-	-	1	2	-	-	-	3
Small Grains									
La Joya Antigua		-	2	5	5	-	-	-	12
Chichas		-	2	2	-	-	-	-	4
Potatoes									
La Joya Antigua		-	7	8	2	-	-	-	17
Chichas		-	1	2	-	-	-	-	3
Onions									
La Joya Antigua		-	-	1	-	1	-	2	4
Chichas		-	-	-	-	2	-	-	2
Other Vegetables									
La Joya Antigua		-	-	-	-	-	1	4	5
Chichas		-	-	-	-	-	1	-	1

TABLE 30 (continued)

Location	Number and Per Cent of Fields								
	Number of Months								Total
	1	2	3	4	5	6	7	8	
La Joya Antigua									
Total	-	2	10	29	31	10	1	6	89
Percent	-	2.2	11.2	32.6	34.8	11.2	1.1	6.7	
Chichas									
Total	-	---	3	9	4	2	1	---	19
Percent	-	---	14.8	47.4	21.1	10.5	5.3	---	
Zamácola									
Total	-	---	---	1	2	---	---	---	3
Percent	-	---	---	33.3	66.6	---	---	---	
Grand Total	-	2	13	39	37	12	2	6	111
Overall Percent	-	1.8	11.7	35.1	33.3	10.8	1.8	5.4	

SOURCE: Field Survey

farmers in the Chichas area to irrigate a particular annual crop type for a shorter number of months than in La Joya Antigua. For example, there is a greater propensity for irrigators to schedule waterings in 4 months in the Chichas area and 5 months in La Joya Antigua (table 30). This situation presents a paradox in that the warmer pampa has a shorter vegetative cycle than that for a similar annual crop type in the Chichas area. The lavish use of irrigation water by the La Joya Antigua irrigator probably reflects an adjustment to sandy soils with little ability to retain moisture.

Frequency and number of irrigations. There are various ways of analyzing the frequency of irrigations. Table 31 depicts several types of frequencies for the crops surveyed in the study area. These methods include the average frequency, the time between planting and the succeeding irrigation, and the time between the last irrigation and the harvest.

There exists a distinct pattern within zones in the average frequency of waterings. In general, regardless of the crop type, the frequency of irrigations is similar. In La Joya Antigua this averages between 6 and 7 days. The only two exceptions are other vegetables and alfalfa. Customarily, throughout the study area, the tradition is to extend the time between waterings for alfalfa beyond that of other crops.

Also the average frequency of waterings in the Chichas area is similar among different crop types, being from 15 to 16 days. There are some exceptions, though, with potatoes and other vegetables (garlic) being irrigated more frequently.

The general pattern in frequency of waterings for these two areas stems from the general frequency in the delivery of water to farms. In La Joya Antigua it is 3 days and in the Chichas area it is from 15 to 16 days. Therefore, in La Joya Antigua, farmers have a tendency to irrigate their crops every other turn whereas in the Chichas area every turn.

The data for the time between planting and the first succeeding irrigation substantiate previously discussed data in which certain crops, as newly set out onions and newly sown alfalfa, if not watered during planting, receive water the next day. On the other hand, most other crops receive their first watering after planting far beyond the average interval between waterings. In part, where the interval is great between planting and the succeeding irrigation, the cause may stem from a heavy pre-planting

TABLE 31

FREQUENCY OF IRRIGATION OF CROPS, CHILI BASIN, 1965-1967

Crop Type and Location	Number of Fields Considered	Days Between Events		
		Average Frequency of Irrigation	Average Interval Between Planting and Succeeding Irrigation	Average Interval Between Irrigation and Harvest
Corn				
La Joya Antigua	51	7	12	7
Chichas	9	15	17	18
Zamácola	3	9	24	14
Small Grains				
La Joya Antigua	12	6	12	3
Chichas	4	16	30	17
Potatoes				
La Joya Antigua	17	7	24	4
Chichas	3	11	24	15
Onions				
La Joya Antigua	4	6	0	7
Chichas	2	15	0	5
Other Vegetables				
La Joya Antigua	5	5	3[a]	6
Chichas	1	12	10	2[b]

TABLE 31 (continued)

Crop Type and Location	Number of Fields Considered	Days Between Events		
		Average Frequency of Irrigation	Average Interval Between Planting and Succeeding Irrigation	Average Interval Between Irrigation and Harvest
Alfalfa				
La Joya Antigua	73	9	1	[c]
Chichas	3	17	1	3

SOURCE: Field Survey

[a]Tomatoes

[b]Garlic

[c]Varies (see text)

watering (aniego or collupo) occurring just before the sowing.

On comparing this parameter of frequency of irrigations between zones, the crops in La Joya Antigua are irrigated within a shorter time after planting than those in the Chichas area. This, in part, reflects the ability of the soils in the latter area to retain more moisture than those in La Joya Antigua.

Data showing the interval of time between the last irrigation and the time of harvest indicate quantitatively the reliance farmers have on the final irrigation in harvesting certain crops (table 31). Customarily onions and garlic are irrigated within a few days before the harvest. Also, at least, in the Chichas area, farmers irrigate alfalfa about 3 days before allowing their livestock to graze it. In La Joya Antigua, this practice is not followed; here irrigations occur in conjunction with the grazing of alfalfa. Farmers in this area notice that little damage results from a compacted wet sandy soil. In fact, during the grazing period, there may be from 1 to 3 waterings applied to a field being grazed.

The range in the number of waterings for a crop type is great, varying from 6 to over 25 (table 32). Crops in La Joya Antigua receive a greater number of waterings than those found in Chichas. More than 40 percent of the fields in La Joya Antigua receive 21 or more waterings, while in Chichas the greatest percentage occurs between 16 and 20 waterings. This variation is not so much a result of a difference in the growing season between the two areas as in the frequency water is delivered to the farmer.

Information on alfalfa is not tabulated in table 32. Never-the less, within an equal time period, it can be deduced from table 31 that a greater number of waterings occur in La Joya Antigua than in Chichas.

Summary. The irrigation season is influenced by the physical environment with the availability of water and the occurrence of frost being important factors. Alfalfa is one of the few crops of the study area that can withstand the effects of both variables. This makes it an ideal crop for the region.

Due to a slightly warmer climate on the Pampa of La Joya than in the Campiña, the irrigation season for an annual crop type is correspondingly shorter in Chichas than in La Joya Antigua. Farmers long ago developed their irrigation season based on decisions made under conditions of evaluation. Now some of the current

TABLE 32

NUMBER OF IRRIGATIONS, BY CROPS, CHILI BASIN, 1965-1967

Crop Type and Location	Number of Fields				
	Ten or Less Irrigations	Eleven to Fifteen Irrigations	Sixteen to Twenty Irrigations	Twenty-one or More Irrigations	Total
Corn					
La Joya Antigua	5	6	23	22	56
Chichas	-	4	--	--	4
Zamácola	-	-	3	--	3
Small Grains					
La Joya Antigua	-	2	5	5	12
Chichas	3	1	--	--	4
Potatoes					
La Joya Antigua	-	3	9	5	17
Chichas	1	2	--	--	3
Onions					
La Joya Antigua	-	-	--	4	4
Chichas	-	2	--	--	2
Other Vegetables					
La Joya Antigua	-	-	--	5	5
Chichas	-	-	1	--	1
Total	9	20	41	41	111

SOURCE: Field Survey

farming decisions of the committed type facing farmers include the timing of planting, harvesting, and irrigating. Now that since a set pattern in the irrigation season has become imbedded in the farming traditions of the people, one wonders what changes will occur from the impact of a more stable and secure water supply brought about by the new large dams and interbasin transfers.

The most significant variable affecting the frequency of waterings throughout the Chili Basin is the interval of time between the delivery of water to a farm. Also, cultural practices unrelated to crop water needs affect the frequency of waterings.

Quantity of Water Employed in Irrigating

Obtaining data on the amounts of water applied by farmers to crops in the study area involved elaborate field observations and measurements. Partly, in conjunction with recording data on scheduling of waterings, 38 fields were included in this portion of the Arequipa irrigation study. Some of the information collected appears in table 33.

Flows used in irrigating. The volume of water flowing to farms in the study area depends principally on the amount of water in the Chili River and on the surreptitious use of it upstream. Due to the variability in stream flow, a farmer faces an uncertain situation on the actual amount to expect at his farm turnout. This condition affects the smooth scheduling of waterings; the ordering of waterings; the hiring of day-laborers, and, even the land-use pattern.

The average stream flow handled in irrigating a field usually equals the flow at the farm turnout. In some instances when one irrigator handles the first watering of a row crop or alfalfa only a part of the total flow entering the farm turnout is employed in one field; the remaining flow goes to another field unattended.

A difference occurs among zones in the amount of water used in irrigating (table 33). The flow entering fields in La Joya Antigua averages 129 l/s, in the Zamácola Project 117 l/s, and in the Chichas area 94 l/s. These flows are similar in quantity to the 100 l/s planned for San Isidro. Also, variations occur in flows within zones. There occurs a greater range in the Chichas area than in either Zamácola or La Joya Antigua. Using average values per irrigation, the Chichas area shows a range from 49 to

TABLE 33

IRRIGATION WATER APPLICATION RECORD FROM SELECTED FIELDS IN THE CAMPIÑA AND LA JOYA ANTIGUA

Zone, Crop and Number of Fields	Average Rate of Flow Used per Watering (l/s)	Depth of water Applied (cm)		Average Amount of Tailwater to Total Volume Applied per Watering (%)	Man-Hours in Watering per Ha	
		Total per Annual	Average per Watering		Total per Annual	Average per Watering
Chichas Area Corn (8)	110	127.2	12.0	4.6	63.5	5.9
Zamácola Corn (3)	117	312.7	19.5	a	121.3	7.6
La Joya Antigua Corn (8)	129	179.3	8.1	a	a	a
Chichas Area Alfalfa (2)	94	-----	13.4	6.3	-----	6.0
La Joya Antigua Alfalfa (5)	140	-----	8.4	a	-----	3.7
Chichas Area Wheat (3)	74	126.9	16.0	9.2	76.5	9.3
Chichas Area Barley (1)	80	109.2	15.6	1.1	60.5	8.6
Chichas Area Potatoes (1)	49	164.9	18.3	10.6	101.1	11.2
La Joya Antigua Potatoes (3)	113	141.5	7.2	a	a	a
Chichas Area Onions (1)	101	211.2	17.6	5.2	65.6	5.5

TABLE 33 (continued)

La Joya Antigua Onions (2)	124	164.6	7.1	a	a	a
Chichas Area Garlic (1)	78	233.2	13.0	12.8	124.8	6.9

SOURCE: Field Survey.

[a]Not measured

168 l/s; the Zamácola Project, from 103 to 132 l/s; and La Joya Antigua from 80 to 184 l/s. Location and water rights of a particular zone affect the range in flows. The Zamácola Irrigation Project with the smallest range, 29 l/s, has a guaranteed allotment of water and its headgate, located along the upper end of the Chili River within the Campiña, allows it to receive a relatively constant flow year-round. On the other hand, the Chichas Canal while guaranteed a specific percentage of water is situated downstream from a series of canals along the Chili River within the Campiña. This results in the Chichas Canal being at the mercy of these other canals in the amount of water permitted to flow into its headgate. La Joya Antigua, also, experiences a great range in the amount of water flowing to farms. Being at the lower end of a river system where many canals above it divert water, the La Joya Antigua farmer can expect great fluctuations in water entering his farm turnouts.

All other factors being constant, the greater the discharge employed, the less time it takes to complete an irrigation (figure 35). To a degree this relationship is more evident in the Campiña than in La Joya Antigua. Differences in the texture of soils may be an important factor affecting this regional variation. Soils are light in La Joya Antigua and correspondingly the infiltration rates are great. Therefore, even with large flows moving into fields, the movement of water down slope is not as rapid as on the heavier soils of the Chichas area.

Depth of water applied. The amount of water applied affects the outcome of the irrigation process. The application of a large depth, wetting the soil well below the root zone of a mature crop, results in a waste of water in the immediate area.

A great range occurs in the total depth applied for different crop types as well as for the same crop type within an area. For example, in the study area, to obtain a corn crop, the total depth applied ranges from 70 to over 300 cm. For different crop types within the same zone, the range is from 100 to 260 cm. In part, this variation is related to the total number of waterings scheduled (figure 36). The greater the total number of waterings, the greater the total depth applied.

As indicated above, for the same crop type, there is a greater number of waterings occurring in La Joya Antigua than

Fig. 35

RELATION BETWEEN THE AVERAGE DISCHARGE AND THE AVERAGE DURATION OF A WATERING PER HECTARE

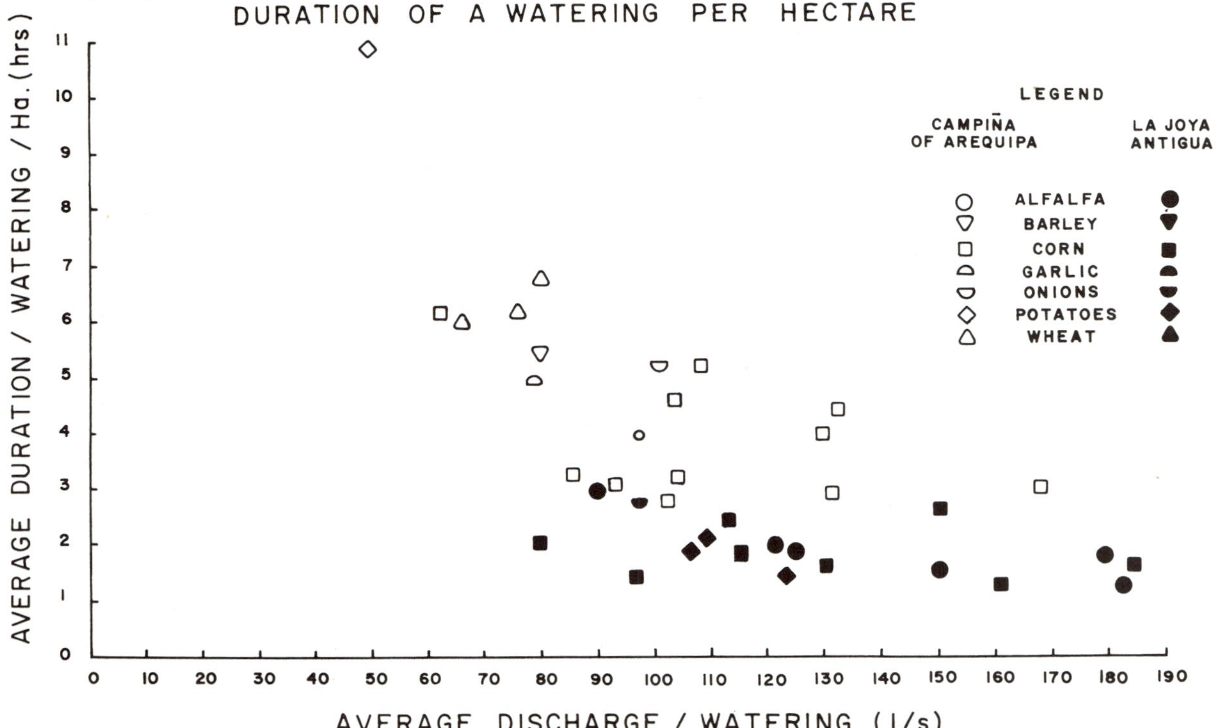

Source : Field Survey .

Fig. 36

RELATION BETWEEN THE TOTAL NUMBER OF WATERINGS AND THE TOTAL DEPTH OF WATER APPLIED TO ANNUALS

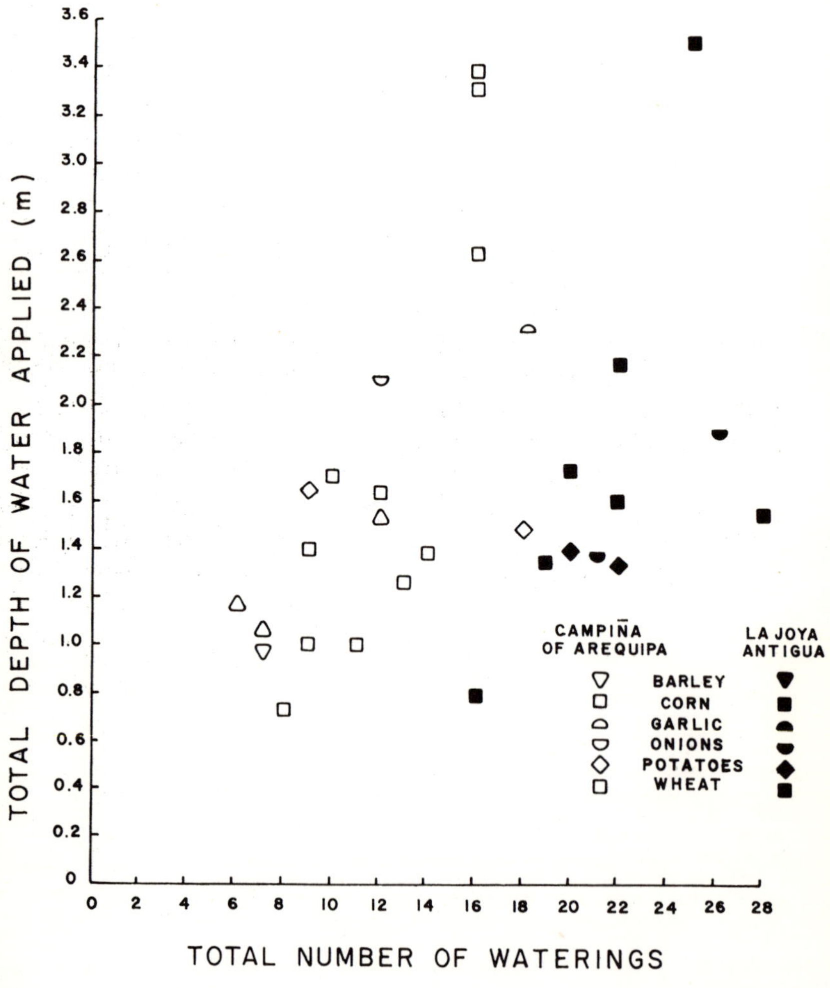

Source : Field Survey .

in the Campiña. As a consequence, in some instances, for similar crop types, a greater total depth of water is applied in the former area than in the latter area. Corn on the average receives a total depth of 180 cm in La Joya Antigua while in the Chichas area it receives 127 cm.

The average depth applied per irrigation in the Campiña is almost double (13 cm) the average in La Joya Antigua (8 cm). Since the soils in the Campiña in general have a larger capacity to store water than in La Joya Antigua, this general relation is expected, but this does not imply that the appropriate amounts are being applied.

As already indicated, it is important to irrigate frequently at the beginning of the vegetative cycle for some crops, until the plants establish themselves. At these times only light irrigations should be applied, certainly no more than 5 cm. However, this does not occur. For some fields in the Campiña the depth applied during these first waterings varies from 8 to 20 cm, while in La Joya Antigua the depth approximates 10 cm.

Further evidence shows that the depth applied does not always relate to the depth of the root zone (figure 37). In many instances a shallow-rooted crop receives a greater depth than a deep-rooted crop. This is especially evident in comparing alfalfa with most other crops. It seems that farmers, believing that the more water applied the greater the yield, have carried this dictum to the extreme in irrigating some of their cash crops. Crops are not irrigated when water is always needed and need is not determined by local field conditions.

Tailwater. There are many factors affecting the amount of tailwater (runoff) from an irrigation. The more important ones include: the longitudinal slope of the field; the infiltration rate, including the wetness of the soil at the time of irrigation; the control of the water as it is applied to the field; and the amount of water being handled by an irrigator. At times the irrigator does not try to control the amount of tailwater. This action stems from a lack of sufficient help in irrigating, especially in handling a large stream flow, or from a feeling that the irrigator does not actually consider tailwater a loss since it is used to wet an adjacent field down slope. Considering the Chili Basin as a unit, tailwater really is not a loss for the area since most of it eventually flows back into the river system,

Fig. 37

RELATION BETWEEN THE AVERAGE DEPTH OF THE ROOT ZONE OF MATURE CROPS AND THE AVERAGE DEPTH OF WATER APPLIED PER WATERING

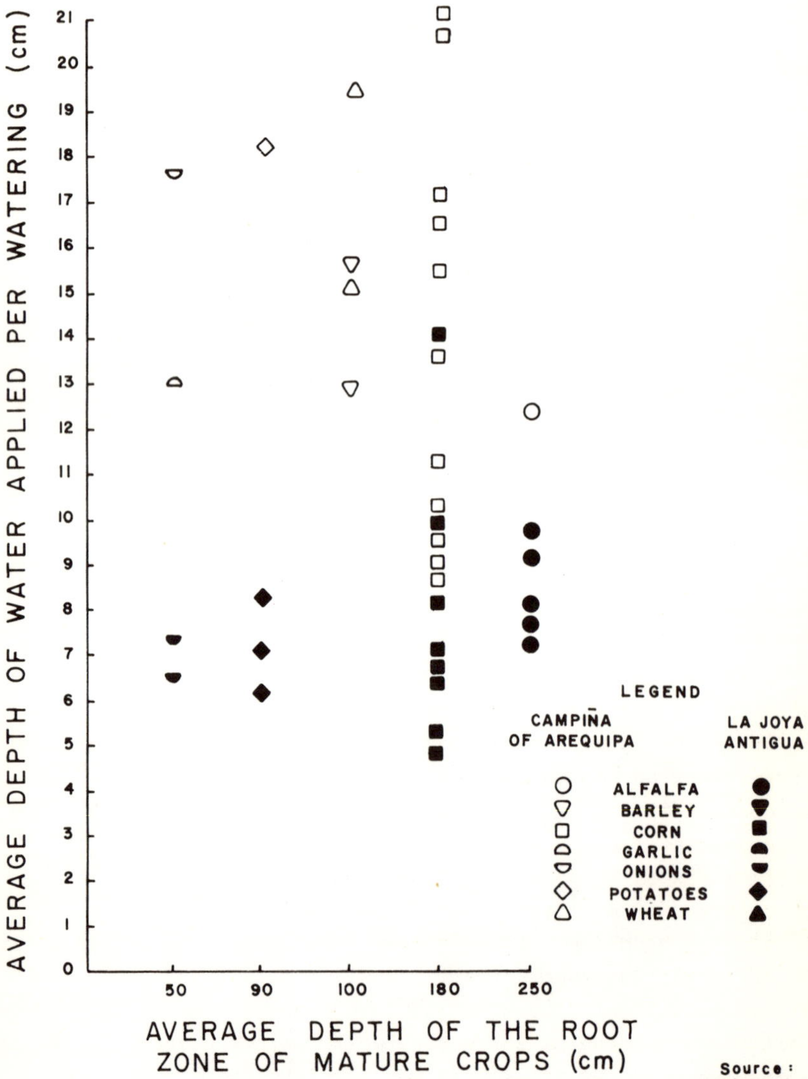

Source: Field Survey.

probably at a lower quality, to be used by farmers further along the stream.

Information from 17 fields in the Chichas area indicates the amount of tailwater resulting from irrigations. There is a great range in the amount of tailwater from one watering to the next. The amount of water applied that flows off a field as tailwater ranges from zero to 14 percent. In many instances during the beginning of the vegetative cycle the amount of tailwater ranges from non-existent to slight. At other times, especially when the crops are mature and the foliage blocks the irrigator's view of the movement of the water, tailwater may account for 25 percent of the total water applied to the field.

A unique condition causing more than the normal amount of tailwater relates to the conditions of the furrows following the previous irrigation. At times, if the water in a furrow flows more slowly than the water in an adjacent furrow, the irrigator in order to speed the flow of water in the first furrow cuts an opening through the ridge separating the two furrows and places a shovel-full of soil in the wetted furrow just downstream from the opening (figure 38). This action diverts the water, and equalizes the flow. If farmers do not clear the clogged furrows before the next watering, a definite unevenness of flow results leading to an appreciable amount of tailwater.

Another time when much tailwater can occur is during the irrigation that precedes the laying out of the irrigation method. During this irrigation (<u>aniego</u>) the amount of tailwater can become as great as 18 percent of the total water applied; yet it has been observed that at other times during an <u>aniego</u> no tailwater occurs. In applying water during an <u>aniego</u>, irrigators are very busy turning over the soil with a shovel in order to have the water penetrate as deeply as possible. They do not seem to have the time to worry about excess water flowing off the field. Also, during <u>aniegos</u>, irrigators customarily use large stream flows. With only one or two irrigators and the extra work involved in distributing water throughout the field, an appreciable amount of tailwater will drain off the field. The depth of water also seems to affect the amount of tailwater. From the limited data available there appears to exist a slight tendency for the amount of tailwater to increase with an increase in the depth applied.

Fig. 38. Applying water to a newly planted garlic field. To redirect more water to furrows with little water, the irrigator cuts gaps in ridges and places small dams across adjacent furrows with excess water.

Stream employed in irrigating furrows. The stream of water applied to a furrow should be of sufficient flow to arrive at the foot of a field without causing appreciable erosion. Faced with this dilemma, farmers have a difficult decision to make in arriving at an optimum choice.

Two ends of this decision-making spectrum are seen in the Chili Basin. Generally at the beginning of the vegetative cycle, irrigators use small streams. During the first watering after the planting of garlic, irrigators usually employ a small stream in order to avoid erosion of land or plants. But in all probability, appreciable amounts of water loss occur through deep percolation at the head of the field. During other irrigations, irrigators at times employ large streams that result in noticeable soil erosion.

There exist differences in this irrigation parameter on a regional basis. The average range in the stream applied per furrow is greater in the Campiña (4 to 12 l/s) than in La Joya Antigua (3 to 5 l/s). The variations in the stream applied per furrow is influenced by the nature of the irrigation method employed. The layout of the irrigation method, particularly the arrangement of the mudas (head furrows leading to the main furrows) affect the ability of an irrigator to handle the flow of water at his disposal. With combinations of 5 furrows for a muda, there is a limit in the precision of an irrigator in controlling the flow of water into each furrow, not withstanding the fact that the chief irrigator, camayo, and his helpers work intensively at applying water. Possibly it is the uneven field surface coupled with many odd length furrows that cause more difficulty in regulating stream sizes in the Campiña than in La Joya Antigua. This difficulty in regulating stream furrow size accounts in part for their variations in the two areas.

Comparing the actual furrow stream sizes used in the study area with suggested sizes from the technical literature indicates a great difference between reality and theory. According to some experts,[1] there is no direct method for determining the proper

---

[1] A. Alvine Bishop et al., "Surface Irrigation Systems," in Irrigation of Agricultural Lands, p. 883.

size of furrow stream. There are many variables involved, including: changes in time of intake rates, advance rates, erodibility, and crop requirements. Experimental methods have been devised to determine information about the impact of these variables on the furrow stream size, but these experiments are both time consuming and require well qualified personnel.[1]

Nevertheless, there are some general technical guides that can be used to aid in determining the maximum stream size or at least the amount that should be permitted. One empirically derived guide (U.S.D.A. model) states that the maximum stream that can be used to avoid soil erosion is equal to 10/S, where S represents the percent slope of the furrow.[2] Another model is that proposed by Zimmerman (table 34).[3] In line with what was previously stated, Zimmerman indicates that the information of the type in table 34 can only be used as an approximate model, the real values being determined from actual field experimentation. But carefully conducted field experiments have not been realized in the Chili Basin.

If these models can be accepted as general guidelines, then the irrigators in the study area, in many instances, use exceptionally large streams per furrow. According to the U.S.D.A. model, the furrow stream should not be any greater than 1 l/s in the study area. Comparing Zimmerman's model to probably the only field that has furrows somewhat circular and wide, the field of garlic in the Chichas area, the irrigator should not apply more than 2 l/s. Actually after the first irrigation, he applies between 3 and 10 l/s per furrow.

In general, irrigators permit more than 4 times the water to flow through a furrow than that recommended by technical studies. While farmers appear to be cautious in using small stream furrow sizes at the beginning of a new crop season, they soon relax their control and permit large streams to flow through furrows.

---

[1] Wayne D. Criddle et al., Methods for Evaluating Irrigation Systems, United States Department of Agriculture, Soil Conservation Service, Agriculture Handbook no. 82 (Washington, D.C.: Government Printing Office, 1956).

[2] John T. Phelan and Wayne D. Criddle, "Surface Irrigation Methods," in Water, The Yearbook of Agriculture, The United States Department of Agriculture (Washington, D.C.: Government Printing Office, 1955), p. 262.

[3] Zimmerman, Irrigation, p. 116.

TABLE 34

MAXIMUM NONEROSIVE UNIT "RUN-IN" STREAM FOR WIDE-CIRCULAR CONSOLIDATED FURROWS AND CORRUGATIONS

Per Cent Grade	0.5	0.8	1.0	1.5	2.0	3.0	4.0	5.0	6.0	7.0	8.0
Flow Rate per Furrow (l/s)	2.6	2.0	1.7	1.1	0.9	0.6	0.4	0.3	0.3	0.2	0.2

SOURCE: Zimmerman, Irrigation, p. 116.

Number of irrigators employed in watering. While the quality and quantity of labor is important to the operation of a farm, it is most important in irrigated agriculture. The task of irrigating adds an extra burden to the farm operation. There exist many factors that influence the number of irrigators employed in applying water. Some of these variables are: the amount of water available, the irrigation method employed, and the availability of irrigators.

Another important factor influencing the watering process especially in the Chili Basin is the traditions of the large number of farm laborers used as irrigators who come from the Altiplano of Puno. Many of these Indians or cholos bring with them a work ethic and habits which benefit the area. The Serrano (Indian and Cholo) is willing to do hard work and, as readily observable, works intensively with crops and fields giving the appearance of a horticultural operation.

A difference exists between zones in the number of irrigators employed per irrigation. In the Campiña more than half of the irrigations have only 1 irrigator handling the application of water, while in La Joya Antigua in the majority of instances there are 3 irrigators. One of the principal reasons for this regional difference is that the latter area usually irrigates with a larger volume of water than in the Campiña (table 33). Therefore, in order to effectively handle these large flows and avoid the rapid erosion of borders and ridges, it is prudent for a greater number of irrigators to be employed. On the other hand, during the first irrigation after the planting of onions, garlic, alfalfa, and corn more irrigators are used. In the Campiña during this crucial time from 2 to 5 irrigators are employed while in La Joya Antigua from 3 to 6 irrigators work.

Another way of expressing the difference between zones in labor input in irrigating is the number of man-hours expended to irrigate a ha (man-hour/watering/ha). On the average 7 man-hours/watering/ha is expended in the Campiña to only 4 man-hours/watering/ha in La Joya Antigua. This major regional difference seems to be due largely to the large stream flows utilized in the latter area. There appears to be an indirect relation between the average discharge available and the man-hour/watering/ha. (figure 39).

On occasions the man-hours/watering/ha ranges from 14 to 25. These conditions occur during the first irrigation for the more delicate crops when many irrigators work and less than 100 l/s enter a field.

Just as there exists a range in the amount of labor expended in irrigating between zones, the same occurs within areas. For example, in the Chichas area to irrigate corn the total man-hours ranges from 40 to 130. The most important variable appears to be the total number of waterings available.

Summary. Farmers in the study area have had to live with uncertainty in terms of the amount of water to expect at their farm turnout. But canals with secure water rights and headgates located far up river have dependable water supplies.

Normally the entire flow received at the farm turnout is used to irrigate one field at a time. However, when initially watering newly planted row crops and alfalfa, only a portion of the total flow to a farm may be used.

There exists a regional difference in the amount of water received by a farmer. Irrigators in La Joya Antigua use, on the average, a greater flow than irrigators in the Chichas area.

A great range exists in the depth of water applied between and within areas. The need for irrigating is not determined by local field conditions. Irrigators often apply a greater depth to shallow-rooted crops than to deep-rooted ones. Farmers, being reluctant to skimp on watering onions, a valuable cash crop, have a tendency to over-irrigate it. The nature of the irrigation method as in the wild flooding of alfalfa, leads to great variations in the depths applied throughout the year.

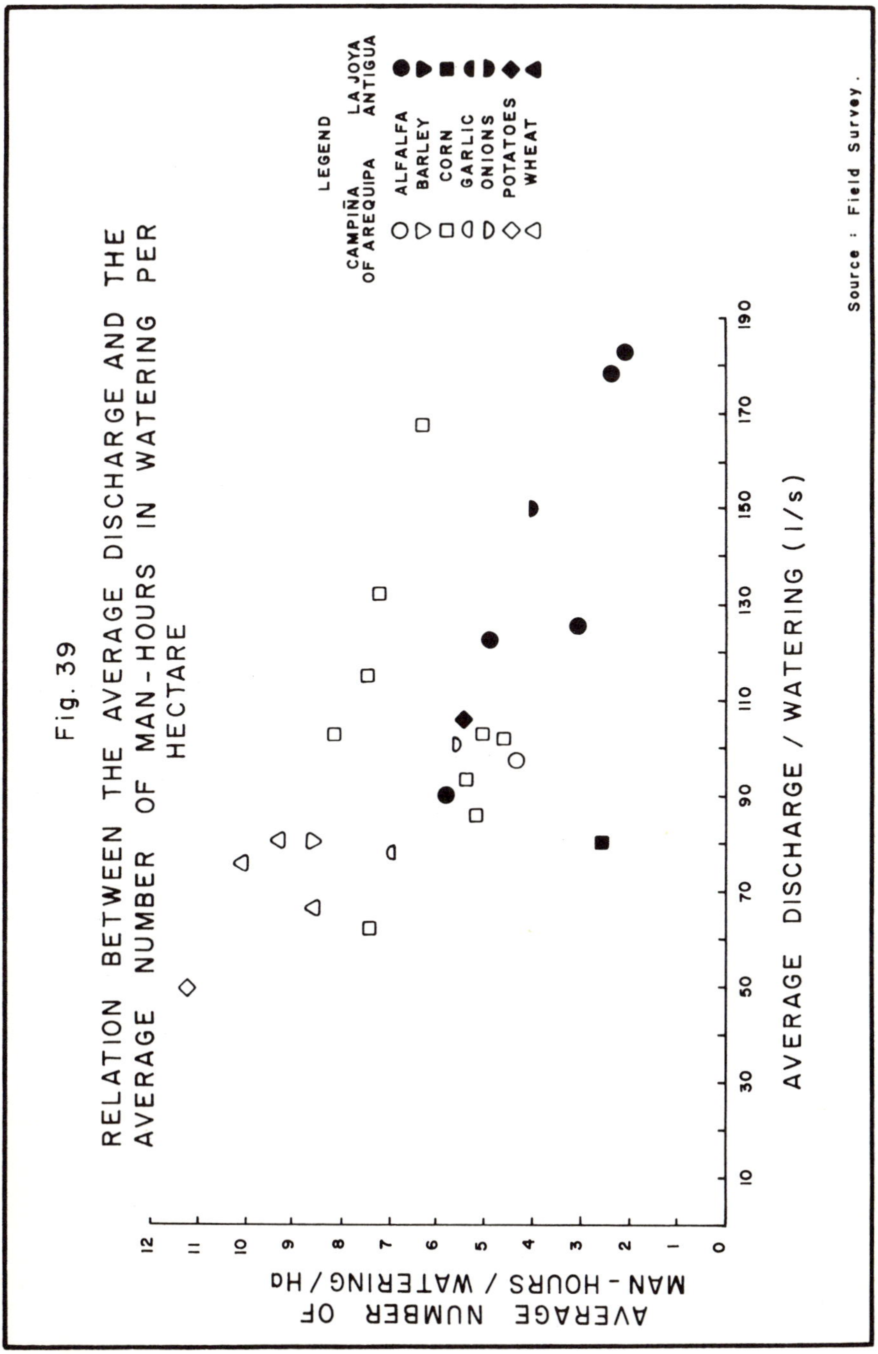

Fig. 39

RELATION BETWEEN THE AVERAGE DISCHARGE AND THE AVERAGE NUMBER OF MAN-HOURS IN WATERING PER HECTARE

The amount of tailwater flowing off fields in the study area varies from one irrigation to the next. Several factors controlling these flows include the volume of water being applied in relation to the number of irrigators working, and the ability of the irrigator to observe what is happening to the flow as it advances down field.

While modifying the alignment of furrows during a watering by an irrigator may increase the amount of tailwater at the next irrigation, the layout of the irrigation method may also have a tendency to lessen runoff. The maze of terminal furrows or border strips forces the water at the lower end of the field to zig-zag back and forth at the bottom of the field providing time for water to penetrate the soil before it flows from the field.

Stream furrow size employed in the study area varies greatly between the first irrigation and the succeeding ones. To a degree the stream furrow size used is influenced by the design of the irrigation method. The arrangement of <u>mudas</u> leads to an automatic funneling of water into a prescribed number of furrows or border strips simultaneously. Variations in the stream furrow size between areas is affected by the nature of the soils and the uniformity in the length of furrows.

There occurs a difference in the ideal stream furrow size as proposed in the literature and that used by farmers in the study area. Not being able to conduct the needed experiments to arrive at a more optimum solution to this phase of the watering process, farmers have settled for an over-use of water.

The number of irrigators used in watering varies throughout the life of the crop. During many of the initial waterings of a crop, many more irrigators are employed than at later waterings in order to use safely in one field all the water received at the farm turnout.

Regional differences exist in the number of irrigators employed per watering. Since a larger volume of water is normally used in La Joya Antigua than in Chichas, the former area employs a greater number of irrigators simultaneously. Also, throughout the study area there exists an indirect relation between the average discharge available and the man-hours used in irrigating. In addition, there occurs a great variation among areas in the total number of man-hours used for irrigations. This variation is greatly influenced by the frequency of water deliveries to a farm. The greater the frequency of water deliveries, the greater the number of man-hours devoted to watering.

Transferring Water Resources Knowledge in San Isidro

For many years the only irrigation existing in the Chili Basin has been in the vicinity of the City of Arequipa and along the flood plain of the river downstream within the canyon which crosses the Pampa of La Joya. In 1940 the Peruvian government began on this Pampa the first major irrigation project (La Joya Antigua) originally consisting of 3400 hectares.

More recently in the 1960s this earlier government project was extended by 1,000 hectares. Each farm in this new colonization (San Isidro) consists of 10 hectares with provision for the farmer to live on his land. In this manner, the project was designed so that the owner and his immediate family could do most of the farm work including irrigation, thus breaking away from the traditional use of hired labor to help with watering and other farm tasks.

The place of water management in the organization of the colonization process. The several tasks of colonizing the new project were accomplished by several government departments with specific goals. These departments consisted of the economic, the agricultural, the cooperative and the water resource sections.

The economic section focused on farm feasibility studies, obtaining and distributing funds to farmers as well as aiding them in the development of homes. The agricultural section primarily supplied technical aid on inputs pertaining to amounts and varieties of crops, fertilizers, and pesticides. In addition, they were in charge of the livestock program including the construction of corrals and the dairying enterprise. The marketing system was handled largely by a cooperative section to which all farmers belonged.

The technical branch on water resources had its task divided into two principal units: the preparation of land for irrigation and the administration of water resources. The possible adverse effects of having these two tasks performed by the same personnel is commented on below.

Since all sectors of the technical staff were under the pressure of a fixed time schedule, it was found that this element seemed to influence greatly the decision-making process. This was especially evident in the amount and kind of attention given to the establishment of a viable on-farm water management program. It is instructive to examine that effort in the light of what is known about the prevailing irrigation practices in neighboring areas.

Specific water management techniques introduced at the initiation of colonization. In making improvements concerning existing practices as found in the general area, technicians introduced several innovations that were to a degree new to the farmers. The three principal parts of on-farm water management affected were the farm water delivery system, the irrigation methods, and the manner of conducting water to fields.

On-farm ditch system. Water received at the farm turnout was distributed through the ten hectare plots by lined ditches all of which served as head ditches. Cementing the farm water distribution system was a great change over the widespread unlined farm ditches used in neighboring areas. The nature and location of this ditching was designed by the technicians taking into consideration the lay of the land as well as the type and characteristics of the irrigation methods to be employed on the new farms.

Irrigation methods. Since the customary practice followed in establishing newly farmed land is to plant a legume such as alfalfa, the technicians decided to institute border-strip irrigation. Their intentions were to make this a somewhat permanent form of irrigation throughout the life of this crop which might be from two to four years. As already referred to, time was ever on the minds of the planners and this had an adverse effect in not allowing them to give sufficient time to properly design the border-strips. A combination of experience in the development of irrigation methods together with classical textbook knowledge and some cursory local experimentation played a role in designing the border-strips.

Water transmission from head ditch to field. With concrete head ditches siphons were used to deliver water from the head ditch to the fields. Various sizes of plastic siphons were used and all farmers were supplied with a set. All were advised as to how to use this equipment through mass demonstration as well as individually.

How and why farmers departed from the water management techniques newly introduced. At first all seemed to be going well with the introduction of these innovations in managing irrigation water but within a few short weeks after the commencement of watering, setbacks began to occur. First, holes were punctured in the walls of the concrete-lined head ditches. Second, the use of siphons was abandoned and finally the border-strip irrigation was abandoned for wild flooding.

Irrigation methods. Remembering that the technicians had planned for border-strip irrigation to be a semipermanent way of spreading water over the fields, it was most discouraging to see their efforts vanish. To a large degree this change in irrigation methods was beyond the control of the planner. Since farmers were unable to obtain the financial resources promised them in establishing corrals for their livestock so that this would allow them to bring cut alfalfa to the penned animals, the farmers were obliged to take their cattle to the field as was the custom. It is interesting to speculate what might have happened had the farmers been provided with corrals. Would there still have been a breakdown in the cattle-alfalfa pattern of feeding? In any case, since it was the customary practice to tether their cattle in the fields, there was no real incentive for farmers to push for corrals. Likewise, the use of hired laborers and their families who lived on the farm, gave an incentive to keep them busy at all times and one of the traditional tasks especially for their children was to care for the tethered cattle in the fields of alfalfa. Thus, with the first "cutting" of alfalfa by the grazing cattle, farmers rather than keeping the banks of the border-strips in good repair (figure 40) or keeping the cattle off them, resorted to their next regular step in irrigating alfalfa, that of wild flooding.

Water transmission from head ditch to field. Long before the abandonment of border-strips for wild flooding, farmers began experiencing difficulties with the use of siphons. Trouble stemmed more from the nature of the physical characteristics of the prepared land and the cultural background of the people than ill-prepared farmers. With the necessity for short runs in the sandy soils, the time of an irrigation setting was rather brief, making

Fig. 40. Partly destroyed borders of border strips in San Isidro. This occurred during the initial grazing of the alfalfa.

for constant changes of settings which frustrated farmers unaccustomed to such work. In many instances this would compel farmers to have two irrigators working simultaneously, one near the head of the field and the other distributing water throughout the middle and lower portions of the border-strip. Another facet making for the necessity of constant and close vigil of water movement to the field from the head ditch was the need to adjust the flow of water through the system of siphons in accordance with the variations in flows coming into the supply ditch. If this equilibrium between inflow and outflow was not maintained, it could and did on a number of occasions result in the erosion of the berm of head ditches and cause their eventual destruction.

Among some farmers there was even the complaint that due to the great head of water entering the field, a serious erosion problem was created, leading to the washing out of young alfalfa plants and leaving large areas vacant. This did not occur under the traditional systems in use where both the head of water was better controlled and the layout of the field under border-strip irrigation allowed for small borders (mudas) to parallel the head ditch allowing for a more complete use of the upper portion of a border-strip.

Since water delivery was on a rotational basis with alternating turns occurring at night, special difficulties were encountered in irrigating in the dark. For proper irrigation to take place at this time, great care in watering was needed. In actuality, a noticeable decrease in the effective control of water occurred at this time. Following the traditional practice, fewer irrigators were used at night. Also among irrigators, night irrigation is looked upon as a less stringent time for watering crops, a time when irrigators tend to relax and irrigate fields that can tolerate less careful scrutiny without gross negative results taking place. Thus irrigators at night throughout the older projects tend to concentrate on large fields of alfalfa or small grains. But these conditions did not exist in the new project. Old habits persisted and devastating results occurred during these night irrigations with damaged head ditches and large amounts of runoff from the fields occurring. Another complaint heard from the night irrigators was the need for very frequent contacts with the cold water with their hands so as to manipulate the siphons on winter nights. Each of these complications built up to a crescendo preparing the way for bolting from the use of siphons for the more traditional delivery system of getting water from the head ditch to the fields.

Once siphons were abandoned, farmers under the supervision of the technical staff perforated small holes on both walls of the head ditch so as to direct water to the adjacent fields (figure 41). An opening was so located that it would serve to irrigate two adjacent border-strips (figure 42). But this undertaking was not sufficient for the purpose. By locating the openings at one edge of the border-strip and in addition to there being small openings, farmers had difficulty getting a sufficient head to irrigate effectively. It appeared that engineers by trying to economize, with small openings seemingly strategically located, allowed finances to dominate their final decision.

When wild flooding replaced border-strips the problem of water delivery from head ditch to field escalated. Now with the definite need for a large head appropriately located for wild flooding, these initial small openings definitely were insufficient. And so the final stage in the transfer of water resource knowledge to the farmer took place; the complete takeover from the change agent of on-farm water management by the farmer himself.

In returning full cycle to their traditional manner of irrigation, farmers without the aid of the technician, selected the location and size of a new set of openings in the walls of the head ditches. In most cases they plugged up the original small holes, and so located the much larger openings as to be positioned next to the newly constructed field ditch running diagonally down the field that now was irrigated by wild flooding (figure 43). In this final stage, the technicians, apparently not willing to condone further deviations from their original irrigation plan, abandoned the farmer.

Fig. 41. Openings (boquetes) in head ditch in San Isidro. Two types of boquetes are shown representing two stages in the evolution of this system. The plugged opening in the background represents the first attempt.

Fig. 42. Irrigating by border strip in San Isidro. One opening in the head ditch is used to irrigate two strips. In the distance is shown barchan dunes on the edge of this irrigation scheme.

Fig. 43. Wild flooding in San Isidro. In the foreground is the large new field ditch from which water flows over the field. Remnants of the old abandoned borders of the original method of irrigation, border strips, can be seen in the background.

CHAPTER V

THE DYNAMICS OF IRRIGATION WATER USE IN THE CHILI BASIN

Irrigated agriculture has been practiced in the Chili Basin for many years, and traditions about this form of agriculture have become deeply embedded in the local culture. These traditions may have had great value in organizing attitudes and practices toward irrigated agriculture in the area, but in recent years, with the development of large dams and interbasin transfers of water, new opportunities have arisen for modifying existing irrigation systems and developing new ones, and traditional practices may not be compatible with new conditions. Many farmers in the newly irrigated areas will come from the surrounding areas, especially the Campiña of Arequipa and other existing irrigated regions in the Chili Basin. These farmers can be expected to bring with them their own somewhat different traditions of managing water. These different traditions will meet in the same areas, and new conditions will influence the application of these traditions as irrigation agriculture expands to meet new opportunities.

In attempting a fuller understanding of the ever changing scene concerning irrigated agriculture in the Chili Basin, it is worth asking how irrigators differ from each other in their management of water, how they depart from what the "tecnicos" regard as optimal, and how the two sets of differences are accounted for. The observations in the Chili Basin throw light on the extent of, as well as the reasons for, differences that exist. Lessons learned from this case study are relevant not only for Peru, but for other areas of the world which must deal with irrigated agriculture as changes occur.

Variations between Farmers' Water Management Practices

Significant differences in the manner in which farmers manage irrigation water occur between and within areas. Although there are differences, a disjointed progression from old to new irrigation projects still exists.

Farmers differ in their management of irrigation water as

shown in the conveyance system developed, their knowledge and perception of the water supply, the methods used in watering, the nature of scheduling waterings, the amount of water applied, and the manner of accepting water resource knowledge in a new irrigation project. These differences reflect the farmers' adjustment to the physical environment within the limits of their traditions and the manner in which the government has intervened in this sector of the economy.

Generally, within each area, the conveyance system is similar. The conveyance system in the Chichas and La Joya Antigua areas shows signs of wear. There has been some maintenance of the system by both farmers and government. The methods used by farmers in diverting water to their farms are much alike throughout the area. Since the system in San Isidro was so new at the time of the study, obviously little difference as yet existed here.

There are, however, significant differences between zones in the conveyance system. Such variables as the density, the symmetry, and even the slope of canals differ. In the Chichas area, a greater array of ditches has evolved, and this phenomenon also is beginning to occur in La Joya Antigua. Whereas engineers, in planning and developing conveyance systems for the projects on the Pampa of La Joya, have been able to control the slopes of the main canal, they have had difficulties in developing the proper slope for laterals, necessitating immediate modifications upon colonizing San Isidro.

The knowledge and perception of the water supply are relatively uniform in the area. Farmers are especially aware of variations in the quality of water they receive.

Notable differences exist in the irrigation methods between the areas in the Chili Basin. Where a common set of irrigation methods of similar design has evolved in the established irrigated areas of the Campiña and La Joya Antigua, a distinct break has occurred in the initial design of the irrigation methods in San Isidro.

In scheduling water most farmers rely upon the changes in color and wilting of foliage. Similar crop types are irrigated for fewer months in the Chichas area than in La Joya Antigua. Although there are major differences in the average frequencies of watering between zones, there is, surprisingly, little difference within zones even among different crop types. A common pattern has developed between planting and the first watering among crop types. For the more delicate crops such as onions and alfalfa, the interval between planting and first watering is short. In La Joya Antigua, the interval is noticeably shorter. Prior to harvesting onions and garlic, whether or not there is a plant water need, the farmer usually irrigates. The number of waterings varies among crop types within the same area as well as between areas. Crops in La Joya Antigua receive more waterings than those of a similar type in the Chichas area.

The amount of water received by a farmer displays a definite spatial pattern. Farmers in La Joya Antigua receive on the average 130 l/s, ranging from 80-184 l/s, whereas the Chichas farmers receive 94 l/s, with a range of 49-168 l/s. Farmers in the Zamácola Irrigation Project receive on the average 117 l/s and experience the smallest range of the three, 103-132 l/s. Of course the greater the discharge employed, the less time it takes to irrigate, but this relationship is more noticeable in the Chichas area than in La Joya Antigua.

Crops are not necessarily irrigated when water is needed, nor is need determined by local field conditions. In many instances farmers apply a greater depth to shallow-rooted crops than to deep-rooted ones. The depth of water applied also varies with the zone. In La Joya Antigua, crops receive an average of 8 cm per irrigation, but in the Chichas area the average is 13 cm.

Tailwater varies from irrigation to irrigation. During the pre-planting waterings, tailwater can amount to as much as 14 percent of the total volume applied. On the other hand, during the first watering after planting, there is often no run-off.

Stream furrow size varies between areas. In the Chichas area the range is 4-12 l/s, whereas the range in La Joya Antigua is from 3-5 l/s. Finally, another variable between zones is the number of irrigators employed. Again there is a noticeable regional difference. On the average, farmers in La Joya Antigua employe three irrigators, but in the Chichas area it is not uncommon to find only one irrigator during waterings. This regional difference in the number of irrigators exists also for the first waterings. La Joya Antigua employs, at this time, from 3-6 irrigators, while the Chichas area employs from 2-5. There is an indirect relationship between the average discharge and the average number of man-hours in irrigating. A regional difference also occurs. In the Chichas area, on the average, 7 man-hours are utilized per irrigation, whereas in La Joya Antigua 4 man-hours are utilized. Variations within a zone pertaining to this parameter exist. For example, corn requires between 40-130 man-hours in the Chichas area.

Many Farmers Depart from What "Technicos" Regard as Optimal

Many farmers use more water than experts estimate is required for crop growth. Engineers in the area tend to introduce high-level irrigation technology, but farmers rely more on simple irrigation methods. Wild flooding is commonly used throughout the study area, yet the practice is viewed by irrigation engineers as an inadequate means of distributing irrigation water. Although engineers did not advocate the abandonment of surface irrigation in the study area at the time of the first expansion of the La Joya Project, they recommended sprinkler irrigation be used at later stages. During this first extension of the La Joya Project irrigation techniques, such as siphons and redesigned border strips, unfamiliar to the local farmers, were introduced. For a short time, the farmers tried using this exotic irrigation technology, but in less than one year they abandoned it for their accustomed practices.

Explanation of Variations among Farmers in Managing Their Water Resources

Irrigation methods vary between the established irrigation projects and San Isidro due principally to unilateral action on the part of the government. However, some of the variance in the design of irrigation methods in the established irrigated areas stem from the physical environment. The rugged and highly terraced Chichas area presents problems and opportunities unlike those encountered on the Pampa of La Joya. These projects show that man is limited in modifying the physical environment to develop a suitable irrigation system. Since it is convenient, farmers rely on plant color and wilting for the scheduling of waterings. Farmers in La Joya Antigua irrigate similar crop types for a longer period of time than farmers in the Chichas area do. This reflects a more lavish use of water on the Pampa of La Joya where the soils are sandy and do not readily retain water. The frequency in the delivery of water to the farms largely determines the frequency of waterings within a region.

Variations in the amount of water received by a farmer is affected by the main canal's location within the river basin. Being located at the head of the basin is advantageous since few farmers can surreptitiously divert water and deprive downstream farmers of their full water rights. Thus, the Zamácola Irrigation Project, located at the head of the Chili Basin, experiences only a small fluctuation in the amount of water received by farmers.

Due to the great infiltration rate of soils in La Joya Antigua, only minor decreases in time to irrigate are realized with the greater discharge available. In addition, the greater the number of waterings, the greater the total depth applied per crop type. Since soils in the Chichas area retain more moisture than soils in La Joya Antigua, farmers in the former area have tended to apply a greater depth of water per watering. Because farmers in La Joya Antigua use larger volumes of water in irrigating, they correspondingly use more irrigators during a watering than is used in Chichas, better to control the irrigation process.

Explanation of Variations Between Farmers and The Technical-Scientific Community in Managing Resources

In general, two factors lead to the differences in the ways farmers and the technical-scientific community manage water and land resources in irrigated agriculture: attitudes toward uncertainty and differing intensities of focus and concern.

Farmers tend to manage their resources to minimize their maximum losses, whereas the expert focuses on maximizing this gains. As soon as farmers realize a new irrigation technique might not work, they fall back on traditional methods. The expert, on confronting resistance to change, is more apt to modify the system rather than allow farmers to abandon it.

Although the irrigation extension agent is almost completely concerned about the proper use of irrigation water, the farmer has a much broader view of the situation. Farmers consider irrigation water management as one among several inputs to successful farming. Engineers under the pressure of a time schedule may not give adequate attention to designing the proper irrigation techniques or to disseminating them properly so they can be used readily by the farmer. Some experts dispense technical aid about irrigation management practices, regulate water distribution, and levy water fines, thereby creating an uneasy feeling among farmers in receiving this aid.

This case study makes clear that the transfer of water knowledge is a difficult process, with societal and cultural ramifications. It is much more than a simple engineering endeavor. The manner in which resource managers, be they farmers or techicians, perceive and act upon the several variables that influence a final decision in the use of a resource is of fundamental concern. Are the time element and the economic factors to be given overriding weight, or do the restraints of past farming practices and present farming convenience take precedence over all else? By allowing the same person to act as irrigation extension agent and administrator in regulating the distribution of water, do project developers run the risk of farmers not wholeheartedly accepting technical aid? Likewise, in arriving at a decision, should project managers make allowance for such uncertainties as the unavailability of expected financial help?

Two radically different alternatives might be considered by the technical-scientific community in improving the use of irrigation water: (1) use the most advanced irrigation system available, and (2) encourage the farmer to design the best irrigation system which he would judge as the most practical system to use.

Even though some areas of the world are in need of simple, cheap, and easily maintained irrigation systems, a possible alternative in trying to improve irrigated agriculture is to break completely with the past where irrigation mangement practices are concerned. By going only part way with a modified gravity system, the farmer has the opportunity easily to revert to his accustomed irrigation methods if problems arise. But, with a more expensive automated system, such a drip irrigation, it might be more difficult for farmers to abandon the system--if for no other reason, than that the authorities would rush to the aid of farmers in order to salvage a heavily funded operation.

Another alternative to be considered in making improvements in the management of irrigation water is to obtain ideas from farmers by making a detailed survey on how they perceive and manage their resources prior to designing a system. This approach should result in a shared responsibility between farmers and experts, reversing the tendency for changes to move from expert to farmer rather than the reverse. An additional benefit received from this approach would expose what parts of an irrigation system farmers perceive worthy of keeping and which parts they think should be abandoned and replaced by more efficient methods. This alternative under somewhat different circumstances, has been tried with success in the Vicos Project in Peru.

Policy Implications

From these observations, experience with irrigation studies in other parts of Peru, and reports of appraisals of irrigation practice elsewhere, two simple and fundamental lessons can be drawn.

1. Farmers adjust to the situation of water availability in a number of ingenious ways which may result in mixed blessings to man and his environment. As a consequence of this mixture more effective measures need to be adopted, particularly in the redesign of surface irrigation methods. In order for meaningful adoptions to occur, however, farmers will require governmental services in the areas of research, information, and education.

2. Engineers tend to explain unsatisfactory performance of a scheme in terms of the "backwardness" of the farmers themselves. To counteract this simplistic view, project planners should determine how and why irrigators now operate before they design improved systems. In taking note of the successes and failures of the past, planners must be cognizant of the underlying factors behind the lay-out of the conveyance network, the farmers' knowledge and perception of the water supply, the methods used in watering, the nature of scheduling irrigations, the amounts of water applied, and the manner of transferring water resource knowledge.

An analysis of existing conditions prior to design would enable planners to develop programs in accord with the traditions and tested practices of the area. Without such action these programs are unlikely to generate widespread acceptance of the new system by the irrigators.

The major policy implication derived from this study is that the technical-scientific community should recognize the essential rationality of traditional farmers, who have a deep reservoir of useful knowledge about their physical environment. A comprehensive and people-oriented irrigation policy should be implemented. For planning purposes, Peru should take steps to raise the status of the farmer to a position approaching that of the technician. The central principle of that policy is that the small farmer, with guidance from the government, must play a vital and active role in rejuvenating irrigation in Peru.

As an essential prerequisite of a people-oriented irrigation policy, Peru must develop within the technical services a cadre of social science specialists such as sociologists, anthropologists, and geographers. Generally, the planning and designing of irrigation schemes are exclusively done by civil engineers and, to a lesser extent, by agronomists. The planning process is thus biased against input from the farming community and favors the structural approach to the solution of problems. Only a new cadre of social science analysts committed to social as well as technical change can reverse this practice.

AUTHOR'S POSTSCRIPT

Since the termination of this study, new irrigation projects have come into existence in the Arequipa Region. Expansion of the La Joya Irrigation Scheme has added 7,000 hectares, and the Majes Irrigation Project, where colonization began in 1982, will eventually add 57,000 hectares.

Farmers in these new projects must now use sprinkler irrigation as the sole means for applying water to their land. This new technology truly represents a clear break from former practices. To aid farmers in using this and other advanced agricultural technologies, particularly within the newly created schemes, a joint agricultural project sponsored by the governments of Israel, the Netherlands, and Peru was created. A tripartite team of engineers and scientists functioned between 1979 and 1982 at the San Camilo Agricultural Experiment Station on the Pampa of La Joya. During this period, they conducted experiments including some with drip irrigation and developed practical irrigation procedures that farmers could use in sprinkler irrigation.[1] These experiments and extension efforts will continue under the guidance of Peruvian engineers who worked with this international team. In addition, the overseas countries plan periodically to send some of their engineers and scientists to Arequipa to continue the technical linkage. As a consequence, this vital and active international link may help to sustain a viable and effectively managed irrigation system among small farmers for many years.

While the herculean task of introducing a radically different method of applying irrigation water to land (sprinkling) represents a worthy venture, only a limited number of farmers will benefit. The majority of irrigators in the Arequipa Region and, in fact, in Peru still rely on surface methods. Since most farmers will not have access to sprinkling, a major scientific-technical thrust to improve surface methods for applying irrigation water to the land is still needed.

---

[1]Isaac Zipor, Mark N. Versteeg, and L. Huber Valdivia, eds., Factores que Influyen en la Producción de Cultivos Alimenticios y Forrajeros en Areas Desérticos (Arequipa, Perú: CINIPA, 1982).

SELECTED BIBLIOGRAPHY

Adams, Richard N., et al. Social Change in Latin America Today: Its Implication for United States Policy. New York: Vintage Books, 1960.

Anderson, Raymond L., and Maass, Arthur. A Simulation of Irrigation Systems. U.S. Department of Agriculture. Economic Research Service, Technical Bulletin No. 1431. Washington, D.C.: Government Printing Office, 1974.

Askochensky, A.N. "Basic Trends and Methods of Water Control in the Arid Zones of the Soviet Union." The Problems of the Arid Zone. Paris: Uniesco, Arid Zone Research No. XVIII, 1962.

Benavente, Victor. Historia Sintéctica de Arequipa. n.p., 1940.

Bielorai, H. "Prediction of Irrigation Needs." Arid Zone Irrigation. Edited by B. Yaron, E. Danfors, and Y. Vaadia. New York: Springer-Verlag, 1973.

Bishop, Alvine A., et al. "Surface Irrigation Systems." Irrigation of Agricultural Lands. Edited by Robert M. Hagan et al. Madison, Wisconsin: American Society of Agronomy, 1967.

Biswas, Asit K., and Durie, Robert W. "Sociological Aspects of Water Development." Water Resources Bulletin, VII (December, 1971), 1137-1143.

Booher, L.J. Surface Irrigation. Rome: Food and Agriculture Organization, 1967.

Bos, M.G., and Nugteren, J. On Irrigation Efficiencies. Wageningen, The Netherlands: International Institute for Land Reclamation and Improvement, 1974.

Boulding, Kenneth E. "Social Sciences." The Great Ideas Today. Edited by Robert M. Hutchins and Mortimer J. Adler. Chicago: Encyclopedia Britannica, 1965.

Brunhes, Jean. Human Geography. Chicago: Rand McNally, 1920.

________. L'Irrigation dans la Peninsule Iberique et dans l'Afrique du Nord. Paris, 1902.

Burton, Ian; Kates, Robert; and White, Gilbert F. The Human Ecology of Extreme Geophysical Events. Toronto: University of Toronto, Department of Geography, Working Paper No. 1, 1968.

Cano, Guillermo J. Gobierno y Administración de los Recursos Hídricos del Perú. Lima, Perú: Oficina National de Evaluación de Recursos Naturales, 1966.

Cantor, Leonard M. A World Geography of Irrigation. London: Oliver Boyd, 1967.

Caro Saavedra, Benjamin. "Reglamentos de Distribución de las Aguas de Ragadío en los Valles de la Costa." Informe de la Inspección General Técnica de la Dirección General de Aguas de Regadío del Ministerio de Agricultura. Lima, Perú, 1966. (Mimeographed).

Chambers, Robert. Settlement Schemes in Tropical Africa. New York: Praeger Publishers, 1969.

Chanduví-Acuña, Fernando. "A Study of Evaporation and Evapotranspiration in Peru." Unpublished M.S. thesis, Utah State Univ., Logan, Utah, 1969.

Chavez Diaz, Rosendo F. El Problema de Majes y Siguas: Planificación Regional de Irrigación en Arequipa. n.p., 1959.

Christiansen, Jerald E. "Irrigation and Drainage Problems in Peru, Observations and Recommendations." North Carolina State Agricultural Mission to Peru. Logan, Utah, June, 1964. (Mimeographed.)

Clagett, Helen L. A Guide to the Law and Legal Literature of Peru. Latin American Series No. 20. Washington, D.C. Government Printing Office, 1947.

"Código de Aguas: Anacrónico; Causa Injusticia en Reparto." La Prensa (Lima, Perú), January 1, 1963, p. 7.

Comité Peruana de Zonas Aridas. Informe Nacional Sobre Las Zonas Aridas. Lima: Imprenta Equiduplic, 1963.

Cornego T., Arturo. "Resources of Arid South America." Arid Lands in Transition. Edited by Harold E. Dregne. Washington, D.C.: American Association for the Advancement of Science, 1970.

Coutu, Arthur J., and King, Richard A. The Agricultural Development of Peru. New York: Praeger Publishers, 1969.

Coward, E. Walter, Jr. Research Methodology in the Study of Irrigation Organization: A Review of Approaches and Applications. Seminar Report No. 18. New York: The Agricultural Development Council, Inc., 1978.

Craik, Kenneth H. "The Environmental Dispositions of Environmental Decision-Makers." The Annals of the American Academy of Political and Social Science, CCCLXXXIX (May 1970), 87-94.

Criddle, Wayne D., et al. Methods for Evaluating Irrigation Systems. United States Department of Agriculture, Soil Conservation Service, Agriculture Handbook No. 82. Washington, D.C.: Government Printing Office, 1956.

de Rivero, Alberto. Arequipa en Su IV Centenario. n.p. 1940.

Dew, Edward. Politics in the Altiplano: The Dynamics of Change in Rural Peru. Austin: University of Texas Press, 1969.

de Wilde, John C. Experiences with Agricultural Development in Tropical Africa. Baltimore: The Johns Hopkins University Press, 1967.

Dobyns, Henry F.; Doughty, Paul L.; and Lasswell, Harold D., eds. Peasants, Power, and Applied Social Change: Vicos as a Model. Beverly Hills: Sage Publications, 1971.

Doneen, L.D. Irrigation Practice and Water Management. Irrigation and Drainage Paper No. 1. Rome: Food and Agriculture Organization, 1971.

Escobar, Gabriel M. Organización Social y Cultural del Sur del Perú. Serie: Antropologia Social, No. 7. Mexico, D.F.: Instituto Indigenista Interamericano, 1967.

Equipo Agropecuario Peruano-Israel. "Estudio de Suelos, Preliminar, Colonización La Joya." Arequipa, Perú, 1967. (Mimeographed).

________. Proyecto de Planificación, Colonización "La Joya". Arequipa, Perú: Programa Conjunto OEA-Israel, 1967.

Farmer, B.H. Pioneer Peasant Colonization in Ceylon. London: Oxford University Press, 1957.

Farrington, Ian. "Irrigation and Settlement Patterns, Preliminary Research Results from the North Coast of Peru." Irrigation's Impact on Society. Edited by Theodore E. Downing and McGuire Gibson. Anthropological Papers of the University of Arizona, No. 25 Tucson: The University of Arizona Press, 1974.

Fenner, Clarence N. Geología de los Alrededores de Arequipa. Translated by Carlos Nicholson. Arequipa, Perú: Editorial La Colmena, S.A., 1940.

Finkel, H.J. "The Barchans of Southern Peru." Journal of Geology, LXVII (1959), 614-647.

Fried, Jacob. "Social Organization and Personal Security in a Peruvian Hacienda Indian Community: Vicos." American Anthropologist, LXIV (August, 1962), 771-780.

Gaitskell, Arthur. Gezira: A Story of Development in the Sudan. London: Faber and Faber, 1959.

Gavilán, Marcelino. Análisis y Características de la Nueva Ley General de Aguas del Perú. Rome: Food and Agricultural Organization, September 5, 1969.

Gillin, John P. "Some Signposts for Policy." Social Change in Latin America Today. Edited by Richard N. Adams, et al. New York: Vintage Books, 1960.

Haise, Howard R., and Hagan, Robert M. "Soil, Plant, and Evaporative Measurements as Criteria for Scheduling Irrigation." in Irrigation of Agricultural Lands, Edited by Robert M. Hagen et al. Madison, Wisconsin: American Society of Agronomy Publications, 1967.

Halperin, Rhoda. "Conclusion: A Substantive Approach to Peasant Livelihood." Peasant Livelihood, Studies in Economic Anthropology and Cultural Ecology. Edited by Rhoda Halperin and James Dow. New York: St. Martin's Press, 1977.

Hawley, Arthur J. "The Decision to Irrigate: Irrigation Practices and the Water Supply in the South Plains of Texas." Unpub. Ph.D. dissertation, Department of Geography, University of Michigan, 1968.

Holling, C.S., ed. Adaptive Environmental Assessment and Management. Chichester, England: John Wiley and Sons, 1978.

Holmes, Roland C. "Evaluación de Riegos - Consumos de Aguas." Paper presented at the III Congreso Nacional de Ingenieros Agrónomos, Lima, Perú, 1965.

________. "La Evaluación del Uso del Agua de Regadío en la Campiña de Arequipa y la Irrigación, La Joya y su Ampliación." A report to the Oficina Nacional de Reforma Agraria del Perú, Arequipa Office, March 1966. (Typewritten).

________. "The Transfer of Irrigation Water Knowledge in the Developing World: Irrigation Water Use in the Arequipa Region of Southern Peru." Water Knowledge Transfer. Vol II. Edited by Neil S. Grigg, Proceedings of the Second International Conference on Transfer of Water Resources Knowledge, June, 1977. Fort Collins, Colorado: Water Resources Publications, 1978.

Holmes, Roland C., and Lizaraso C., Benjamín. "Informe Sobre Reconocimiento del Estado Actual de Captación de las Aguas de Rio Chili en la Campiña de Arequipa." A Report to the Director del Programa, La Joya, Arequipa, Peru, 1966. (Typewritten).

Houk, Ivan E. Irrigation Engineering. II vols. New York: John Wiley and Sons, Inc., 1951.

Hudson, James. Irrigation Water Use in the Utah Valley, Utah. Department of Geography Research Paper No. 79. Chicago: University of Chicago, Department of Geography, 1962.

Ingersoll, Jasper. "Anthropologists and the Agency for International Development (A.I.D.): An Old Hate Relationship and a New Love Affair." Anthropological Quarterly, L (October 1977), 199-203.

Inter-American Committee for Agricultural Development. Inventory of Information Basic to the Planning of Agricultural Development in Latin America: Peru. Washington D.C." Pan American Union, 1964.

Israelsen, Orson W. Irrigation Principles and Practices, 2d ed. New York: John Wiley and Sons, Inc., 1950.

Israelsen, Orson W., and Hansen, Vaughn E. Irrigation Principles and Practices, 3d ed. New York: John Wiley and Sons, Inc., 1962.

James, L. Douglas, ed. Man and Water. Lexington, Kentucky: The University Press of Kentucky, 1974.

Jenks, William F. "An Account of the Geology of the Arequipa Region." Peruvian Times, February, 1946, pp. 25-31.

________. Geología de la Hoja de Arequipa al 200,000. Buletín no. 9. Lima, Perú: Instituto Geológico de Perú, 1948.

Johnson, A.M. "The Climate of Peru, Bolivia and Ecuador." Climates of Central and South America. Edited by Werner Schwerdtfeger. Amsterdam: Elsevier Scientific Publishing Company, 1976.

Keller, Jack. "Effect of Irrigation Method on Water Conservation." Journal of the Irrigation and Drainage Division Proceedings of the American Society of Civil Engineers, XCI (June, 1965), 61-72.

Keller, Wesley, and Calson, Car W. "Forage Crops," in Irrigation of Agricultural Lands, Edited by Robert M. Hagan et al. Madison, Wisconsin: American Society of Agronomy, 1967.

Kolars, John F. "Decision and Commitment in Turkish Agriculture." Review of the Geographical Institute. University of Istanbul, International Edition, No. 11 Separate, 1965-1968.

Kovda, V.A., ed. Irrigation, Drainage and Salinity: An International Source Book. London: Hutchinson and Company, 1973.

Kuiper, Edward. Water Resources Development. London: Butterworths, 1965.

Lanning, Edward P. Peru Before the Incas. Englewood Cliffs, New Jersey: Prentice-Hall, Inc., 1967.

"Las Administraciones Técnicas de Aguas de Regadió y la Necesidad de su Subsistencia." Aguas e Irrigación. (Lima, Perú). 1, No. 1. (May 15, 1932), 82-101.

Lasarte, E. "Estudio Relative a la Reglamentación de los Derechos de Agua de Regadió." Aguas e Irrigación (Lima, Perú). 1, No. 2. (June 1932), 11-14.

Lettau, Heinz. "Dynamic and Energetic Factors which Cause and Limit Aridity along South America's Pacific Coast." Climates of Central and South America. Edited by Werner Schwerdtfeger. Amsterdam: Elsevier Scientific Publishing Company, 1976.

Lettau, Katharina, and Lettau, Heinz. "Bulk Transport of Sand by the Barchans of the Pampa de La Joya in Southern Peru." Zeitchrift für Geomorphologie, n.f. XIII (1969), 182-195.

Linsley, Ray K., and Franzini, Joseph B. Water Resources Engineering. New York: McGraw-Hill Book Company, 1964.

Lizárraga, C.A. "El Estudio del Riego." Aguas e Irrigación (Lima, Perú), 1, No. 2 (June, 1932), 41-47.

Lydolph, Paul E. "On the Causes of Aridity along a Selected Group of Coasts." Coastal Deserts, Their Natural and Human Environments. Edited by David H.K. Amiran and Andrew W. Wilson. Tucson: The University of Arizona Press, 1973.

Maass, Arthur, and Anderson, Raymond L. . . . and the Desert Shall Rejoice: Conflict, Growth, and Justice in Arid Environments. Cambridge, Massachusetts: The MIT Press, 1978.

Malpica, Carlos. "El Régimen Legal de las Aguas." Revista de Economía y Agricultura. (Lima, Perú). I (June - August, 1964), 309-313.

Mariátegui, José Carlos. 7 Ensayos de Interpretación de la Realidad Peruana. 3rd ed. Lima, Perú: Empresa Editora Amauta S.A., 1952.

Nicolini, Juan Vicente. "La Policiá de Las Aguas en el Perú." Revista Universitaria de La Universidad Mayor de San Marcos, (Lima, Perú) XV, vol. II, 1920, 241-267.

Noranha, Raymond. "The Anthropologist as Practitioner: The Field of Development AID." Anthropological Quarterly, L (October 1977), 211-216.

Palmer, Gary. "The Ecology of Resettlement Schemes." Human Organization, XXXIII (Fall, 1974), 239-250.

"Peru's Water Reform Law." Peruvian Times, April 10, 1970, pp. 11-13.

Phelan, John T., and Criddle, Wayne D. "Surface Irrigation Methods." Water, The Yearbook of Agriculture. The United States Department of Agriculture. Washington, D.C.: Government Printing Office, 1955.

Prohaska, Federico J. "New Evidence on the Climatic Controls along the Peruvian Coast." Coastal Deserts, Their Natural and Human Environments. Edited by David H. G. Amiran and Andrew W. Wilson. Tucson: The University of Arizona Press, 1973.

Rawitz, E. "Gravity Irrigation." Arid Zone Irrigation. Edited by B. Yaron; E. Danfors; and Y. Vaadia. New York: Springer-Verlag, 1973.

República del Perú. Ley General de Aguas, Decreto Ley No. 17752, 24 de Julio de 1969. Lima, Perú: Empresa Editora del Diario Oficial, "El Peruano."

_______. Ministerio de Agricultura Oficina Sectorial del Planificación Agraria. Inventario de Proyectos para Incrementar en un Millón de Hectáreas: El Area Cultivada con Programa Preliminar de Ejucución y Estimación de Inversiones. Lima: Editorial Servicio de Copias "Fast", 1965.

_______. Ministerio de Fomento y Obras Publicas. Evaluación Economica y Planeamiento Departmento de Arequipa. Lima: Prensa y Publicaciones, 1958.

_______. Ministerio de Hacienda y Comercio. Dirección Nacional de Estadística y Censos. Censos Nacionales de Población, Vivienda y Agropecuario: 1961. Departamento de Arequipa. Lima, 1969.

_______. Oficina Nacional de Evaluación de Recursos Naturales, Inventorio, Evaluación y Uso Racional de Los Recursos Naturales de La Costa: Cuencas de Los Rios Quilca y Tambo. Vols. III. Lima, 1974.

_______. Plan Regional para el Desarrollo del Sur del Perú, Manual de Estadística Regional. Lima: Servicio Cooperativo Peruano Norteamericano de Educación, 1959.

Roder, Wolf. The Savi Valley Irrigation Projects. Department of Geography, Research Paper No. 99. Chicago, University of Chicago, Department of Geography, 1965.

Rodriquez Escobedo, Carlos. Código de Aguas: Concordancias de Legislación y Jurisprudencia (Leyes y Reglamentos de Irrigacion). Arequipa, Perú: Editorial El Deber, S.A. Ltda, 1956.

Schaedel, Richard P. La Demografía y Los Recursos Humanos del Sur del Perú. Serie: Antropologiá Social. No. 8. Mexico: Instituto Indigenista Interamericano, 1967.

Sewell, W.R. Derrick. "Environmental Perceptions and Attitudes of Engineers and Public Health Officials." Environment and Behavior. III (March, 1971), 23-59.

Shultz Wetzel, Otto, et al. Proyecto de Planificación, Colonización "La Joya." Programa Conjunto OEA-Israel, Equipo Agropecuario Peruano Israelí. Arequipa, Perú: Oficina Nacional de Reforma Agraria, 1967.

Sims, John H., and Baumann, Duane D., eds. Human Behavior and the Environment. Chicago: Maaroufa Press, 1974.

Steward, Julian H., and Faron, Louis C. Native Peoples of South America. New York: McGraw-Hill Book Company, Inc., 1959.

Teaf, Howard M., Jr. and Franck, Peter G. Hands Across Frontiers: Case Studies in Technical Cooperation. Ithaca, New York: Cornell University Press, 1955.

Tosi, Joseph A. Jr. Zonas de Vida Natural en el Perú. Boletín Técnico, no. 5. Lima, Perú: Institute Interamericano de Ciencias Agrícolas de la O.E.A., Zona Andina, 1960.

Trewartha, Glenn T. The Earth's Problem Climates. Madison: Univ. of Wisconsin Press. 1961.

United Nations, Economic Commission for Latin America. "Los Recursos Hidraúlicos de America Latina: Perú." E/CN. 12/794, July 31, 1968. (Mimeographed).

United Nations, Food and Agriculture Organization and the International Bank for Reconstruction and Development. The Agricultural Development of Peru, Part I and II. General Report and Detailed Report, Water Utilization. Washington, D.C.: I.B.R.D., 1959.

United States Department of Commerce. Overseas Business Reports. Basic Data on the Economy of Peru. Washington, D.C.: Government Printing Office, 1969.

United States Department of Interior. Bureau of Reclamation and Bureau of Indian Affairs. Report on the Water Conservation Opportunities Study. Washington, D.C.: Government Printing Office, 1978.

United States Soil Conservation Service and United States Bureau of Reclamation. Irrigation on Western Farms. Agriculture Information Bulletin No. 199. Washington, D.C.: Government Printing Office, 1959.

Varallyay, G. "Soil Water Problems Related to Salinity and Alkalinity in Irrigated Lands." Arid Land Irrigation in Developing Countries. Edited by E. Barton Worthington, New York: Pergamon Press, 1977.

Vargas, Luis V. Geología del Cuadrángulo de Arequipa. Boletín no. 24, Lima, Perú: Servicio de Geología y Mineriá, 1970.

Wagner, Philip L. The Human Use of the Earth. Glencoe, Illinois: The Free Press of Glencoe, 1960.

Ward, Barbara, and Dubos, Rene. Only One Earth: The Care and Maintenance of a Small Planet. New York: W.W. Norton and Co., 1972.

Wolf, Eric R. Sons of the Shaking Earth. Chicago: University of Chicago Press, 1959.

Worthington, E. Barton. Arid Land Irrigation in Developing Countries: Environmental Problems and Effects. New York: Pergamon Press, 1977.

Zimmerman, Josef D. Irrigation. New York: John Wiley and Sons, Inc., 1966.

THE UNIVERSITY OF CHICAGO
DEPARTMENT OF GEOGRAPHY
RESEARCH PAPERS (Lithographed, 6×9 inches)

LIST OF TITLES IN PRINT

48. BOXER, BARUCH. *Israeli Shipping and Foreign Trade.* 1957. 162 p.
56. MURPHY, FRANCIS C. *Regulating Flood-Plain Development.* 1958. 216 pp.
62. GINSBURG, NORTON, editor. *Essays on Geography and Economic Development.* 1960. 173 p.
71. GILBERT, EDMUND WILLIAM. *The University Town in England and West Germany.* 1961. 79 p.
72. BOXER, BARUCH. *Ocean Shipping in the Evolution of Hong Kong.* 1961. 108 p.
91. HILL, A. DAVID. *The Changing Landscape of a Mexican Municipio, Villa Las Rosas, Chiapas.* 1964. 121 p.
97. BOWDEN, LEONARD W. *Diffusion of the Decision To Irrigate: Simulation of the Spread of a New Resource Management Practice in the Colorado Northern High Plans.* 1965. 146 pp.
98. KATES, ROBERT W. *Industrial Flood Losses: Damage Estimation in the Lehigh Valley.* 1965. 76 pp.
101. RAY, D. MICHAEL. *Market Potential and Economic Shadow: A Quantitative Analysis of Industrial Location in Southern Ontario.* 1965. 164 p.
102. AHMAD, QAZI. *Indian Cities: Characteristics and Correlates.* 1965. 184 p.
103. BARNUM, H. GARDINER. *Market Centers and Hinterlands in Baden-Württemberg.* 1966. 172 p.
105. SEWELL, W. R. DERRICK, *et al. Human Dimensions of Weather Modification.* 1966. 423 p.
107. SOLZMAN, DAVID M. *Waterway Industrial Sites: A Chicago Case Study.* 1967. 138 p.
108. KASPERSON, ROGER E. *The Dodecanese: Diversity and Unity in Island Politics.* 1967. 184 p.
109. LOWENTHAL, DAVID, editor, *Environmental Perception and Behavior.* 1967. 88 p.
112. BOURNE, LARRY S. *Private Redevelopment of the Central City, Spatial Processes of Structural Change in the City of Toronto.* 1967. 199 p.
113. BRUSH, JOHN E., and GAUTHIER, HOWARD L., JR., *Service Centers and Consumer Trips: Studies on the Philadelphia Metropolitan Fringe.* 1968. 182 p.
114. CLARKSON, JAMES D., *The Cultural Ecology of a Chinese Village: Cameron Highlands, Malaysia.* 1968. 174 p.
115. BURTON, IAN, KATES, ROBERT W., and SNEAD, RODMAN E. *The Human Ecology of Coastal Flood Hazard in Megalopolis.* 1968, 196 p.
117. WONG, SHUE TUCK, *Perception of Choice and Factors Affecting Industrial Water Supply Decisions in Northeastern Illinois.* 1968. 93 p.
118. JOHNSON, DOUGLAS L. *The Nature of Nomadism: A Comparative Study of Pastoral Migrations in Southwestern Asia and Northern Africa.* 1969. 200 p.
119. DIENES, LESLIE. Locational Factors and Locational Developments in the Soviet Chemical Industry. 1969. 262 p.
120. MIHELIČ, DUŠAN. *The Political Element in the Port Geography of Trieste.* 1969. 104 p.
121. BAUMANN, DUANE D. *The Recreational Use of Domestic Water Supply Reservoirs: Perception and Choice.* 1969. 125 p.
122. LIND, AULIS O. *Coastal Landforms of Cat Island, Bahamas: A Study of Holocene Accretionary Topography and Sea-Level Change.* 1969. 156 p.
123. WHITNEY, JOSEPH B. R. *China: Area, Administration and Nation Building.* 1970. 198 p.
124. EARICKSON, ROBERT. *The Spatial Behavior of Hospital Patients: A Behavioral Approach to Spatial Interaction in Metropolitan Chicago.* 1970. 138 p.
125. DAY, JOHN CHADWICK. *Managing the Lower Rio Grande: An Experience in International River Development.* 1970. 274 p.
126. MACIVER, IAN. *Urban Water Supply Alternatives: Perception and Choice in the Grand Basin Ontario.* 1970. 178 p.
127. GOHEEN, PETER G. *Victorian Toronto, 1850 to 1900: Pattern and Process of Growth.* 1970. 278 p.
128. GOOD, CHARLES M. *Rural Markets and Trade in East Africa.* 1970. 252 p.
129. MEYER, DAVID R. *Spatial Variation of Black Urban Households.* 1970. 127 p.
130. GLADFELTER, BRUCE G. *Meseta and Campiña Landforms in Central Spain: A Geomorphology of the Alto Henares Basin.* 1971. 204 p.
131. NEILS, ELAINE M. *Reservation to City: Indian Migration and Federal Relocation.* 1971. 198 p.
132. MOLINE, NORMAN T. *Mobility and the Small Town, 1900–1930.* 1971. 169 p.

133. SCHWIND, PAUL J. *Migration and Regional Development in the United States.* 1971. 170 p.

134. PYLE, GERALD F. *Heart Disease, Cancer and Stroke in Chicago: A Geographical Analysis with Facilities, Plans for 1980.* 1971. 292 p.

135. JOHNSON, JAMES F. *Renovated Waste Water: An Alternative Source of Municipal Water Supply in the United States.* 1971. 155 p.

136. BUTZER, KARL W. *Recent History of an Ethiopian Delta: The Omo River and the Level of Lake Rudolf.* 1971. 184 p.

139. MCMANIS, DOUGLAS R. *European Impressions of the New England Coast, 1497–1620.* 1972. 147 p.

140. COHEN, YEHOSHUA S. *Diffusion of an Innovation in an Urban System: The Spread of Planned Regional Shopping Centers in the United States, 1949–1968,* 1972. 136 p.

141. MITCHELL, NORA. *The Indian Hill-Station: Kodaikanal.* 1972. 199 p.

142. PLATT, RUTHERFORD H. *The Open Space Decision Process: Spatial Allocation of Costs and Benefits.* 1972. 189 p.

143. GOLANT, STEPHEN M. *The Residential Location and Spatial Behavior of the Elderly: A Canadian Example.* 1972. 226 p.

144. PANNELL, CLIFTON W. *T'ai-chung, T'ai-wan: Structure and Function.* 1973. 200 p.

145. LANKFORD, PHILIP M. *Regional Incomes in the United States, 1929–1967: Level, Distribution, Stability, and Growth.* 1972. 137 p.

146. FREEMAN, DONALD B. *International Trade, Migration, and Capital Flows: A Quantitative Analysis of Spatial Economic Interaction.* 1973. 201 p.

147. MYERS, SARAH K. *Language Shift Among Migrants to Lima, Peru.* 1973. 203 p.

148. JOHNSON, DOUGLAS L. *Jabal al-Akhdar, Cyrenaica: An Historical Geography of Settlement and Livelihood.* 1973. 240 p.

149. YEUNG, YUE-MAN. *National Development Policy and Urban Transformation in Singapore: A Study of Public Housing and the Marketing System.* 1973. 204 p.

150. HALL, FRED L. *Location Criteria for High Schools: Student Transportation and Racial Integration.* 1973. 156 p.

151. ROSENBERG, TERRY J. *Residence, Employment, and Mobility of Puerto Ricans in New York City.* 1974. 230 p.

152. MIKESELL, MARVIN W., editor. *Geographers Abroad: Essays on the Problems and Prospects of Research in Foreign Areas.* 1973. 296 p.

153. OSBORN, JAMES F. *Area, Development Policy, and the Middle City in Malaysia.* 1974. 291 p.

154. WACHT, WALTER F. *The Domestic Air Transportation Network of the United States.* 1974. 98 p.

155. BERRY, BRIAN J. L., *et al. Land Use, Urban Form and Environmental Quality.* 1974. 440 p.

156. MITCHELL, JAMES K. *Community Response to Coastal Erosion: Individual and Collective Adjustments to Hazard on the Atlantic Shore.* 1974. 209 p.

157. COOK, GILLIAN P. *Spatial Dynamics of Business Growth in the Witwatersrand.* 1975. 144 p.

159. PYLE, GERALD F. *et al. The Spatial Dynamics of Crime.* 1974. 221 p.

160. MEYER, JUDITH W. *Diffusion of an American Montessori Education.* 1975. 97 p.

161. SCHMID, JAMES A. *Urban Vegetation: A Review and Chicago Case Study.* 1975. 266 p.

162. LAMB, RICHARD F. *Metropolitan Impacts on Rural America.* 1975. 196 p.

163. FEDOR, THOMAS STANLEY. *Patterns of Urban Growth in the Russian Empire during the Nineteenth Century.* 1975. 245 p.

164. HARRIS, CHAUNCY D. *Guide to Geographical Bibliographies and Reference Works in Russian or on the Soviet Union.* 1975. 478 p.

165. JONES, DONALD W. *Migration and Urban Unemployment in Dualistic Economic Development.* 1975. 174 p.

166. BEDNARZ, ROBERT S. *The Effect of Air Pollution on Property Value in Chicago.* 1975. 111 p.

167. HANNEMANN, MANFRED. *The Diffusion of the Reformation in Southwestern Germany, 1518–1534.* 1975. 248 p.

168. SUBLETT, MICHAEL D. *Farmers on the Road. Interfarm Migration and the Farming of Noncontiguous Lands in Three Midwestern Townships. 1939–1969.* 1975. 228 pp.

169. STETZER, DONALD FOSTER. *Special Districts in Cook County: Toward a Geography of Local Government.* 1975. 189 pp.

170. EARLE, CARVILLE V. *The Evolution of a Tidewater Settlement System: All Hallow's Parish, Maryland, 1650–1783.* 1975. 249 pp.

171. SPODEK, HOWARD. *Urban-Rural Integration in Regional Development: A Case Study of Saurashtra, India—1800–1960.* 1976. 156 pp.

172. COHEN, YEHOSHUA S. and BERRY, BRIAN J. L. *Spatial Components of Manufacturing Change.* 1975. 272 pp.

173. HAYES, CHARLES R. *The Dispersed City: The Case of Piedmont, North Carolina.* 1976. 169 pp.

174. CARGO, DOUGLAS B. *Solid Wastes: Factors Influencing Generation Rates.* 1977. 112 pp.

175. GILLARD, QUENTIN. *Incomes and Accessibility. Metropolitan Labor Force Participation, Commuting, and Income Differentials in the United States, 1960–1970.* 1977. 140 pp.

176. MORGAN, DAVID J. *Patterns of Population Distribution: A Residential Preference Model and Its Dynamic.* 1978. 216 pp.

177. STOKES, HOUSTON H.; JONES, DONALD W. and NEUBURGER, HUGH M. *Unemployment and Adjustment in the Labor Market: A Comparison between the Regional and National Responses.* 1975. 135 pp.

179. HARRIS, CHAUNCY D. *Bibliography of Geography. Part I. Introduction to General Aids.* 1976. 288 pp.

180. CARR, CLAUDIA J. *Pastoralism in Crisis. The Dasanetch and their Ethiopian Lands.* 1977. 339 pp.

181. GOODWIN, GARY C. *Cherokees in Transition: A Study of Changing Culture and Environment Prior to 1775.* 1977. 221 pp.

182. KNIGHT, DAVID B. *A Capital for Canada: Conflict and Compromise in the Nineteenth Century.* 1977. 359 pp.

183. HAIGH, MARTIN J. *The Evolution of Slopes on Artificial Landforms: Blaenavon, Gwent.* 1978. 311 pp.

184. FINK, L. DEE. *Listening to the Learner. An Exploratory Study of Personal Meaning in College Geography Courses.* 1977. 200 pp.

185. HELGREN, DAVID M. *Rivers of Diamonds: An Alluvial History of the Lower Vaal Basin.* 1979. 399 pp.

186. BUTZER, KARL W., editor. *Dimensions of Human Geography: Essays on Some Familiar and Neglected Themes.* 1978. 201 pp.

187. MITSUHASHI, SETSUKO. *Japanese Commodity Flows.* 1978. 185 pp.

188. CARIS, SUSAN L. *Community Attitudes toward Pollution.* 1978. 226 pp.

189. REES, PHILIP M. *Residential Patterns in American Cities, 1960.* 1979. 424 pp.

190. KANNE, EDWARD A. *Fresh Food for Nicosia.* 1979. 116 pp.

192. KIRCHNER, JOHN A. *Sugar and Seasonal Labor Migration: The Case of Tucumán, Argentina.* 1980. 158 pp.

193. HARRIS, CHAUNCY D. and FELLMANN, JEROME D. *International List of Geographical Serials, Third Edition, 1980.* 1980. 457 p.

194. HARRIS, CHAUNCY D. *Annotated World List of Selected Current Geographical Serials, Fourth, Edition. 1980.* 1980. 165 p.

195. LEUNG, CHI-KEUNG. *China: Railway Patterns and National Goals.* 1980. 235 p.

196. LEUNG, CHI-KEUNG and GINSBURG, NORTON S., eds. *China: Urbanization and National Development.* 1980. 280 p.

197. DAICHES, SOL. *People in Distress: A Geographical Perspective on Psychological Well-being.* 1981, 199 p.

198. JOHNSON, JOSEPH T. *Location and Trade Theory: Industrial Location, Comparative Advantage, and the Geographic Pattern of Production in the United States.* 1981. 107 p.

199-200. STEVENSON, ARTHUR J. *The New York-Newark Air Freight System.* 1982. 440 p.

201. LICATE, JACK A. *Creation of a Mexican Landscape: Territorial Organization and Settlement in the Eastern Puebla Basin, 1520–1605.* 1981. 143 p.

202. RUDZITIS, GUNDARS. *Residential Location Determinants of the Older Population.* 1982. 117 p.

203. LIANG, ERNEST P. *China: Railways and Agricultural Development, 1875–1935.* 1982. 186 p.

204. DAHMANN, DONALD C. *Locals and Cosmopolitans: Patterns of Spatial Mobility during the Transition from Youth to Early Adulthood.* 1982. 146 p.

205. FOOTE, KENNETH E. *Color in Public Spaces: Toward a Communication-Based Theory of the Urban Built Environment.* 1983. 153 p.

206. HARRIS, CHAUNCY D. *Bibliography of Geography. Part II: Regional. Vol. 1. The United States of America.* 1984. 178 p.

207-208. WHEATLEY, PAUL. *Nāgara and Commandery: Origins of the Southeast Asian Urban Traditions.* 1983. 473 p.

209. SAARINEN, THOMAS F.; SEAMON, DAVID; and SELL, JAMES L., eds. *Environmental Perception and Behavior: An Inventory and Prospect.* 1984. 263 p.

210. WESCOAT, JAMES L., JR. *Integrated Water Development: Water Use and Conservation Practice in Western Colorado.* 1984. 239 p.

211. DEMKO, GEORGE J., and FUCHS, ROLAND J., eds. *Geographical Studies on the Soviet Union: Essays in Honor of Chauncy D. Harris.* 1984. 294 p.

212. HOLMES, ROLAND C. *Irrigation in Southern Peru: The Chili Basin.* 1986. 191 p.

213. EDMONDS, RICHARD L. *Northern Frontiers of Qing China and Tokugawa Japan: A Comparative Study of Frontier Policy.* 1985. 155 p.

214. FREEMAN, DONALD B., and NORCLIFFE, GLEN B. *Rural Enterprise in Kenya: Development and Spatial Organization of the Nonfarm Sector.* 1985. 180 p.

215. COHEN, YEHOSHUA S., and SHINAR, AMNON. *Neighborhoods and Friendship Networks: A Study of Three Residential Neighborhoods in Jerusalem.* 1985. 129 p.

217-218. CONZEN, MICHAEL P., ED. *World Patterns of Modern Urban Change: Essays in Honor of Chauncy D. Harris.* 1986.

219. KOMOGUCHI, YOSHIMI. *Agricultural Systems in the Tamil Nadu: A Case Study of Peruvalanallur Village.* 1986. 171 p.

220. GINSBURG, NORTON; OBORN, JAMES; and BLANK, GRANT. *Geographic Perspectives on the Wealth of Nations.* 1986. 131 p.